Beyond Interdisciplinarity

Beyond Interdisciplinarity

Boundary Work, Communication, and Collaboration

Julie Thompson Klein

OXFORD
UNIVERSITY PRESS

OXFORD
UNIVERSITY PRESS

Oxford University Press is a department of the University of Oxford. It furthers the University's objective of excellence in research, scholarship, and education by publishing worldwide. Oxford is a registered trade mark of Oxford University Press in the UK and certain other countries.

Published in the United States of America by Oxford University Press
198 Madison Avenue, New York, NY 10016, United States of America.

Library of Congress Control Number: 2021935958
ISBN 978-0-19-757115-6 (pbk.)
ISBN 978-0-19-757114-9 (hbk.)

DOI: 10.1093/oso/9780197571149.001.0001

For George, who never hesitates to question a boundary

And for Sarah, who never hesitated to cross a boundary

Contents

Foreword

The only way to achieve the impossible is to believe that it is possible.
Lewis Carroll, Alice in Wonderland (1865)

Every idea, every concept has an origin, a genesis, and a more or less transformative evolution according to uses, contexts, and scientific, linguistic, social, and cultural practices. The concept of interdisciplinarity in the broad sense—which some say is a simple fad or even denigrate it under the pretext of disciplinary excellence while others make it an object of study and a practice in its own right and perfectly legitimate—belongs to this large family of living concepts in science, arts, culture, and society. The book you hold in your hands or under your digital eye is timely. It is remarkable and will be noticed by all researchers, teachers, all people who are directly or indirectly interested not only in interdisciplinarity in all its forms but also in the future of disciplines in a changing world. Interdisciplinarity as a concept, an epistemological posture and a practice is characterized more by its heterogeneity than by theoretical and methodological uniformity. It is not one and indivisible, but plural and shareable beyond disciplinary, national, and international differences. It works on and at the boundaries of disciplines and comes in more or less multi-, inter- and trans-disciplinary forms (*crossdisciplinary work*), while extending its radius of action in and with extra-academic spheres (*cross-sectorial work*).

Interdisciplinarity as an object and knowledge of interdisciplinarity do not submit to essentialism; they are not reducible to a set of strictly necessary and sufficient definitions which would definitively fix what interdisciplinarity is or is not. However, and conversely, they are not reducible to a relativism of the object and of knowledge which would only be understood in specific and contingent contexts. By avoiding the double trap of essentialism and relativism, we are invited here to think about interdisciplinarity in all nuances. Interdisciplinarity in the diversity and extent of its forms is consistent, in the sense that it has long been the subject of academic research which determines its contours, documents practices and informs the scientific community of advances at the frontiers of disciplines. But it is also fluid in the links that interdisciplinarity forges with other concepts and practices of communication, collaboration, learning, and creativity. The middle path that opens between the acquired consistency and the fluidity in the making is the opportunity to situate the work on interdisciplinarity in a broad and rhizomatic perspective. Interdisciplinary and disciplinary communities need this openness more than ever to fully understand the undeniable advances of disciplinary but also interdisciplinary research, to grasp the inter-misunderstandings and to overcome the epistemological, institutional, and cultural obstacles that still prevent reciprocal recognition and valuation of the diversity of scientific practices. Interdisciplinarity, this concept with a thousand faces, the very plasticity of which allows disciplines and boundary work to be included in the same space and time.

"Beyond" interdisciplinarity means in spatial terms what there is history behind and below the concept in a complex and ramified definitional space; in temporal terms it means what there is after with the idea to go beyond and to go further, toward renewed practices of interdisciplinarity and perhaps more provocative paths such as postdisciplinarity or indisciplinarity. We salute here the erudite and experiential attention paid to diversity, evolution, change, and the future of interdisciplinarity. Julie Thompson Klein invites us on an exciting conceptual journey informed by science and practice. Renowned for her support of young scholars, Julie's characteristic generosity and inclusivity shine through in her acknowledgment of the contributions of others as she shares her encyclopedic understanding of interdisciplinarity and what it means to go "beyond." Julie invites us to accompany her on her academic and personal trajectory in the service of interdisciplinarity, epitomized by her scientific rigor and commitment to scholarship and typified by a humility, enthusiasm, and generosity that are acclaimed by friends and colleagues across the globe.

Frédéric Darbellay
Inter- and Transdisciplinarity Unit
Center for Children's Rights Studies
University of Geneva (Valais Campus)

Catherine Lyall
Science, Technology & Innovation Studies
School of Social and Political Science
University of Edinburgh

March 2021

Acknowledgments

This book is the culmination of a nearly career-long investigation into the nature of interdisciplinarity. During the late 1970s and 1980s I began studying the concept out of curiosity. At the time I was a member of a new experimental curriculum founded in 1973 at Wayne State University. It was initially called the University Studies/Weekend College Program, then subsequently elevated to the Department of Interdisciplinary Studies, before being demoted back to program status. My colleagues and I felt beleaguered by skepticism about our mission in an institution that had already terminated a well-known experiment in the history of interdisciplinary education, Monteith College. Doubts about the rigor of interdisciplinarity were also coupled with the stigma of being called a “weekend” college despite meeting other days of the week as well, an academic reward system that favored research over teaching, and a student population of working adults.

Flash forward to 2007, when the University terminated the program. Interdisciplinary research was highly valued by then, not only locally but nationally and internationally as well. Because my original discipline was literary studies I was placed in the English Department, even though I had evolved into a humanities professor with extensive interests in social science and studies of knowledge and of higher education. I was also active in multiple professional networks engaged in inter- and trans-disciplinarity and team science. The generous support of the Vice President for Research, Hilary Ratner, and the Associate Vice President for Research, Gloria Heppner, led to a buyout of half my teaching load in order to promote interdisciplinary development and team science on campus. I remain grateful for their respect and our shared commitment.

In moving beyond that biographical context I run the risk of failing to acknowledge the multitude of individuals who have nurtured and supported my scholarship since the late 1980s. So I cite here only professional organizations that became homes for my work while thanking individuals whose conversations and collaborations directly helped me test ideas in this particular book. Even that narrow slice of time, however, is rich and abundant. The first organization is the Association for Interdisciplinary Studies (AIS). Given my conviction that a scholar’s best reader is someone who offers constructive criticism, even in disagreement, I am especially grateful for ongoing feedback from former AIS presidents Machiel Keestra at the University of Amsterdam and Rick Szostak at the University of Alberta.

The Network for Transdisciplinary Research, known as td-net, has also been a vital international home. It continues to be a space of learning and collaboration in conferences and projects that cross boundaries of disciplines, fields, professions, and sectors of society with the aim of solving complex problems. I am especially grateful to Christian Pohl, who was instrumental in facilitating my appointment as an International Research Affiliate in the Transdisciplinarity Lab (TdLab) of the Department of Environmental Systems Science at ETH-Zurich. I also continue to benefit from collaborating with and the support of Sabine Hoffmann, Group Leader of Transdisciplinary Research at the Swiss Federal Institute of Aquatic Science and Technology known as EAWAG.

TdLab projects led to productive collaborations with others as well. I have gained valuable insights by co-authoring with Dena Fam and Cynthia Mitchell, formerly at the University of Technology Sydney, learning from their expertise in sustainability research and education. Bianca Vienni Baptista in the TdLab at the University of Zurich also continues to be a valued partner in numerous projects, including a special section of the AIS journal *Issues in Interdisciplinary Studies* on crossdisciplinary research and education in Latin America and a forthcoming book on institutionalizing interdisciplinarity with an international array of case studies.

Four additional individuals associated with td-net are ongoing sources of conversations, collaborations, and kindred projects. Gabriele Bammer at the Australian National University has been a steadfast leader of the Integration and Implementation Sciences network, known as i2S, and has generously included me in many efforts on an international scale. Conversations and publications with Frédéric Darbellay and Zoe Moody at the University of Geneva have also nurtured our mutual interests in inter- and transdisciplinarity. And Catherine Lyall at the University of Edinburgh continues to be a valued partner in several projects throughout Europe, in science policy, higher education, and crossdisciplinary research and education.

The International Network for the Science of Team Science, known as INSciTS, has been another home for advancing my understanding of collaborative research. I continue to learn from colleagues in the Network and wider field of team scholarship. In addition to Daniel Stokols at the University of California, Irvine campus, who has always been a generous partner in gathering and reflecting on resources, I especially thank Michael O'Rourke, Director of the Center for Interdisciplinarity at Michigan State University. He was a vital partner when he chaired the leadership team for hosting the 2019

InSciTS conference, and beyond that remains one of my most valued sources of feedback.

Proper credit for illustrations others created appear in the text, but I want to thank here as well the following people for permission to use their work: Dena Fam at the University of Technology Sydney; Elsbeth Spelt at the University of Wageningen; Stephen Fiore at the University of Central Florida; Daniel Stokols at the University of California, Irvine; and Justin Nash at the University of Connecticut. In addition I thank Noha Beydoun for early help locating resources while she was my graduate research assistant at Wayne State University. I also thank Bethany Laursen of Laursen Evaluation and Design, LLC, and Michigan State University for her design work on my original text tables for this book and her own scholarship on integration.

And, finally, I thank my editor at Oxford University Press, Jeremy Lewis. I have received nothing but wise counsel and generous support throughout the process of composing this book. He is an author's dream. I am also grateful to the staff of the Press for enabling the art of Gail Ryder for the book's cover and advertising material. It is a unique expression of relationships and layerings associated with both crossdisciplinary and cross-sector work. Ryder is an independent artist and Associate Professor of Humanities at Siena Heights University.

Glossary

Table A.1 The Glossary is a digest of related concepts that appear across the book. Proper citations for concepts in **bold print** appear in pertinent chapters.

Overarching Terms	
Crossdisciplinary is a composite term for multi-, inter-, and trans-disciplinary approaches. **Cross-sector** is a composite term for bridging academic, governmental, industrial, and community perspectives and expertise.	
Boundaries	
Overview	Boundaries have a dual character since they demarcate and enclose but are also permeable and contingent. Three related concepts also appear across the book.
Dimensions	**Boundary Rhetoric** signifies territorialization and politicization as well as permeation and crossfertilization. **Boundary Objects** mediate differences by creating common foci, enabling communication and cooperation between groups with different epistemologies, values, norms, and aims. **Boundary Agents and Organizations** perform brokering roles across different forms of knowledge, including both academic expertise and stakeholder groups.
Boundary Work	
Overview	**The conceptual framework of this book is a composite of claims, activities, and structures by which individuals and groups work directly and through institutions to create, maintain, break down, and reformulate knowledge units.**
Dimensions	**Acts** of spanning, crossing, and bridging boundaries; **Inclusions** of multiple forms of knowing, including academic, lay, and Indigenous forms; **Processes** of interacting, integrating, and collaborating; **Strategies** of brokering, mediating, and negotiating; **Operations** of demarcating, constructing, and refiguring; **New Relations** of interdependence and **convergence**; **Outcomes** of breaching, transgressing, and transforming.

(*continued*)

Table A.1 *Continued*

Communication and Collaboration	
Overview	Communication entails exchanging individual approaches and building **Interlanguage** in both **Trading Zones** and **Communities of Practice.**
Dimensions	**Pigeon** forms of language are interim tongues, while **Creoles** are first languages of new communities. **Communication** requires **Translation** across different specialist languages but is not a rote process of application and transfer. It also requires **negotiation, dialogue,** and **translation boundary work.** **Collaboration** entails **Boundary Weaving** through **dialogue** in a **communication boundary space** that develops a **cooperation and communication culture.** In order for **Collective Identity** to emerge in a group, **collective communication competence** is crucial, facilitated by **socio-cognitive platforms** and **collaborative interdisciplinary reasoning** for **integration, communicative rationality,** and **communicative action.**
Disciplinarity and Multidisciplinarity	
Overview	Knowledge is classified into differing forms. The dominant organization in the 20th century was disciplinarity, segmenting subjects and objects into separate domains. Multidisciplinary approaches aim to bring them together but typically lack integration and collaboration.
Dimensions	**Disciplinarity** demarcates selected objects, subjects, and methods that reinforce specialization through institutionalizing mechanisms of academic departments, professional societies, meetings, publications, degrees, and a job market. They establish boundaries around epistemic cultures, dividing insiders from outsiders. Yet, they also change historically and crossfertilize. **Multidisciplinary** approaches juxtapose separate disciplinary approaches around a common interest, adding breadth of knowledge and approaches. However, disciplines continue to speak as separate voices in encyclopedic alignment and retain their original identity. Underlying assumptions are not examined, and the status quo remains intact.
Discourses of Interdisciplinarity	
Overview	Patterns of argument shape the way concepts are defined and valued. They create boundaries around particular claims, practices, and structures that create tensions across them. Priorities shift over time, though, and despite continuing tensions discourses intersect.
Dimensions	**Philosophical Discourse** advances epistemology and ontology by exploring the theoretical nature of knowledge and of reality. Associated historically with the quest for unity of knowledge, it expanded with new theories of knowledge and synthetic paradigms.

Table A.1 *Continued*

Discourses of Interdisciplinarity (cont'd)	
	Discourse of Problem Solving is primarily instrumental, driven in large part by scientific, economic, and technological problems. It prioritizes pragmatic needs, application, criteria of reliability, and efficiency over theoretical interests. **Discourse of Critique** interrogates the existing structure of knowledge and education. It is a multilayered form of argument that interrogates instrumentality, the principle of unity, disciplinarity, and even interdisciplinarity while promoting transgression of boundaries and a democratic imperative
Overlaps	Categories are not air-tight, however. Problem solving is currently prioritized over epistemology, but cross-sector work is fostering an **Epistemic of Problem Solving,** a **Critical Epistemology,** as well as **Transdisciplinary** merging of **Socially Robust Knowledge and Metareflection.** Critique also advances a **Transgressive** approach to existing structure and practices with the aim of transforming them. Discourses intersect with other concepts and theories, including **complexity, Mode 2 knowledge production, postnormal science** and **wicked problems, convergence,** and **team science,** as well as **heterarchy.**
Ecology of Spatializing Practices	
Overview	A variety of sites facilitate cutting across demarcations and **Boundary Weaving:** including **Boundary Zones, Engagement Spaces,** and **Transaction Spaces.**
Dimensions	**Spatial Metaphors** conjure images of turf and territory as well as jurisdiction and control. The dominant bounded space in the academy is disciplines. **Organic Metaphors** signify processes of generation, crossfertilization, interrelation, and hybridization akin to processes in nature.
Integration	
Overview	Integration is widely regarded as essential to crossdisciplinary and cross-sector collaboration. However, there is no universal formula, and degrees of interaction differ.
Dimensions	Four fault lines of definition appear: **Linear and Algorithmic** step models differ from **Heuristic and Constructivist** frameworks. Theoreticians and practitioners differ on whether integration is **Cognitive** in nature or factors in **Social and Communicative** aspects. Proponents disagree on whether it is an **Individual** phenomenon or a **Collaborative** one. Emphasis varies on primacy of **Disciplines** versus **Societal Perspectives.**

(continued)

Table A.1 *Continued*

Integration (Cont'd)	
Overlaps	The underlying premise of unification is disputed. **Partial Integration** of specialities, **Contextualized** alternatives, and **Alternative Frameworks** challenge universal theories, while acknowledging differing scales and contexts. In addition, **thematic, product- or problem-oriented types** of integration differ as well as **epistemic, cognitive, social-organizational,** and **communicative levels**.
Interdisciplinarity (ID)	
Overview	Interdisciplinarity connotes integration of data, methods, tools, concepts, theories, and/or perspectives from multiple disciplines or bodies of knowledge in order to answer a question, to solve a problem, or to address a topic or theme that is too broad or complex to be dealt with by one discipline. It is dated to the early 20th century, though expanded into a heterogeneity of practices and forms, ranging from borrowing tools and methods to forming new fields and interdisciplines. A variety of catalysts and contexts also shape outcomes, with differing trajectories and intersections with other fields.
Dimensions	**Instrumental Forms** prioritize economic, technological, and scientific problem solving, often aligned with needs of the marketplace and national defense. **Critical and Reflective Forms** interrogate the existing structure of knowledge and education while raising questions of value and purpose silent in instrumental forms. **Methodological Forms** tend to improve the quality of results, as in borrowing a method or concept to test a hypothesis, to answer a research question, or to help develop a theory. **Theoretical Forms** build a comprehensive conceptual framework or synthesis or foster systematic integration of propositions, models, or analogies across disciplines. **Narrow ID** occurs between disciplines with compatible methods, paradigms, and epistemologies. **Broad or Wide ID** occurs between disciplines with different epistemologies and methods.
Learning	
Overview	Learning is fundamental to collaboration in both crossdisciplinary and cross-sector work.
Dimensions	**Integrative Learning** in interdisciplinary education is grounded historically in a shift from content-based to process-based integration, the concept of **Unifying versus Unification** of existing approaches, and the theory of **Constructivism** and **Reflective Equilibrium**. **Mutual, adaptive, generative, deep, double- and triple-loop learning, and reflexivity** are linked with **collaborative learning** and **situated learning.** **Relationality** and **Transactivity** foster **collective intelligence.**

Table A.1 *Continued*

Learning (Cont'd)	
	Stage models move from **identification and positioning** to **coordination** and the possibility of **transformation**, while factoring in **iteration** and **mediation**. **Competencies** span disciplinary knowledge and integrative and collaborative capacity, including cognitive and social skills as well as ethics, while cultivating intellectual curiosity, flexibility and adaptability, and the ability to absorb information. **Productive Pedagogies** support interdisciplinary learning with a **multiplicative power** that draws on team-based projects and case studies; role-playing, simulations, and gaming; problem-, discovery-, and inquiry-based learning; and field experiences as well as formal didacticism and mentoring. **Transdisciplinary Orientation** combines values, attitudes, beliefs, conceptual skills and knowledge, and behaviors for **interactional expertise.**
Transdisciplinarity (TD)	
Overview	The quest for **Unity of Knowledge** framed initial thinking. Although initially less prominent than interdisciplinarity, this label become more prominent in the late 20th century, expanding underlying meanings of the core term. It also elevated new relationships between science and society while challenging traditional alignment with disciplines.
Dimensions	The first major typology of interdisciplinary approaches cited the exemplar of anthropology as a broad science of humans. However, participants differed on exact meaning: whether it is a particular interlanguage, a higher stage in epistemology of interdisciplinary relations, or social purpose. Subsequently it labeled **Overarching Syntheses** that transcend the narrow scope of disciplinary worldviews, including **general systems theory, post/structuralism, feminist theory, cultural critique, complexity theory, and sustainability.** Meaning broadened further to include a new overarching **paradigm for health and wellness** and **collaborative cross-sector partnerships for sustainability**. The term has also been a descriptor for **synoptic disciplines and broad fields** as well as new **critical paradigms** fostering transgression of existing boundaries. In the 1980s and 1990s three new major connotations appeared: a **new structure of unity informed by the worldview of complexity in science,** a new mode of **knowledge production** that fosters synthetic reconfiguration and recontextualization, and partnerships with stakeholders in public and private spheres for solving complex problems through **co-production of knowledge** and **public engagement.**

Sources: Klein, 2020; Laursen Evaluation and Design, LLC.

Introduction

Beyond Interdisciplinarity

> How might one understand interdisciplinarity less as a unity and more as a field of differences, a multiplicity?
>
> **Barry and Born, 2013, p. 5**

The opening epigraph is the starting point for this book. In countering the widespread belief that certain forms dominate definition of interdisciplinarity, Barry and Born (2013) urged mapping heterogeneity of the concept instead. They called in particular for more genealogies to track path-dependencies and temporalities in both theory and practice. Heterogeneity of activities and forms associated with the concept has led some to liken it to the proverbial Tower of Babel. However, a plurality of terms does not automatically spell cacophony. It requires sorting out similarities and differences in theories and practices. Echoing Sally Aboelela et al. (2007), definitions of interdisciplinarity vary by modes of research and education, their relationship to disciplinarity, degrees of interaction and integration, patterns of communication and exchange, as well as goals, contexts, and outcomes. This book answers Barry and Born's call by mapping the boundary work of activities often linked with, but not entirely encompassed by, the core word *interdisciplinarity*. The Introduction establishes a framework for doing so, followed by a preview of subsequent chapters. The English writer Lewis Carroll (1896) provided a humorous introduction to the underlying question of meaning.

The character Humpty Dumpty, who gained fame for falling off a wall in a popular nursery rhyme, came to fuller life in Carroll's novels about a young girl's adventures in Wonderland. In *Through the Looking Glass*, Humpty Dumpty and Alice debated the meaning of words:

> "When I use a word," Humpty Dumpty said in rather a scornful tone. "It means just what I choose it to mean—neither more or less."

Beyond Interdisciplinarity. Julie Thompson Klein, Oxford University Press. © Oxford University Press 2021.
DOI: 10.1093/oso/9780197571149.003.0001

> "The question is," said Alice, "whether you can make words mean so many different things."
>
> "The question is," said Humpty Dumpty, "which is to be master—that's all." (Lewis Carroll, 1896)

Interdisciplinarity has likewise been the subject of debate over which definition is to be the master. Introducing the updated edition of *The Oxford Handbook of Interdisciplinarity*, Robert Frodeman (2017) reported the term is most commonly used as a portmanteau word for all more-than-disciplinary approaches to knowledge. The term ***interdisciplinarity*** appears so widely, in fact, that some consider the concept meaningless. Dogan and Pahre (1990) even suggested banishing the word altogether. However, Frodeman argued instead for understanding how ambiguities in associated terms have functioned in the political economy of knowledge. In the course of discussion, this book does so while situating the concept in relation to two others. One—disciplinarity—is deemed both its opposite and its complement, while the other—transdisciplinarity (TD)—is of heightened interest today. In addition the book tracks cross-sections with other related concepts in pertinent chapters.

Each concept has its own literature. Gabriele Bammer (2012) suggested trans-disciplinary problem solving, for example, has until recently been largely a separate research stream with its own handbook, design principles, and review of methods. Yet, related interests intersect and the scope of discussion continues to expand beyond efforts to codify separate professional literatures. The same is true of a concept *du jour* examined more fully in Chapter 2, on Discourses. Convergence has become a popular slogan for a variety of efforts to cross boundaries, though influential organizations are laying claim to sanctioned meanings. At the same time, Daniel Stokols cast doubt on Venn diagrams that depict overlaps of convergence with interdisciplinarity and team science. Related concepts appeared earlier in literatures on community-engaged action research and social ecology. Moreover, even though they share some underlying assumptions, they are not equivalent or entirely overlapping. For instance, early notions of convergence were narrower than broader conceptions of transdisciplinarity and collaborative scholarship, confined to STEM fields (pers. comm., 6 July 2019). Understanding intersections, then, is crucial to understanding the current heterogeneity of interdisciplinarity while factoring in insights from literatures on other related topics in this book, including communication, learning, and institutional change. The discipline of linguistics provides a further gloss on the meaning of words.

The Meaning of a Word

Linguists attribute shifts to multiple causes, though four stand out (Ullman, 1962). The first, Pejoration, signals negative connotations, evident in critiques of both the tradition of unity of knowledge and the current heightened priority of problem solving. It is also apparent in attacks on the very concept of interdisciplinarity. The second cause, Amelioration, signals positive associations, including elevation of problem solving over epistemology. The third, Narrowing, marks restricted uses, including association of "interdisciplinary studies" with education rather than research. It is also evident in the crossfire of claims over what constitutes "genuine" inter- and trans-disciplinarity. Scholarship is not a combat sport, but it appears so in the boundary work of definition. Finally, the fourth, Broadening, acknowledges expanded meaning (Klein, 2020).

Broadening prompts this book, though inclusive of negative, positive, and narrow connotations. Any attempt at definition also benefits from etymological sleuthing. Stephen Fiore (2008) tracked crossdisciplinary collaboration to Brozek and Key's observation in 1944 that the interdisciplinary approach was becoming a prominent characteristic of science. David Sill (1996), though, attributed origin even earlier, citing the 1929–1930 annual report of the US-based Social Science Research Council (SSRC). Roberta Frank (1988) tracked origin of the term even earlier, to the mid-1920s. It was shorthand for crossing boundaries of the seven societies of the SSRC, she reported, with the aim of problem-focused applied research. Taking the 1920s as a point of etymological origin, then, interdisciplinarity is nearing a century mark. A widely cited point of origin, however, does not freeze meaning. During the same period, the underlying idea was associated with integrative models of general education and the core curriculum movement. Deeper into the 20th century problem-solving underwent Amelioration, expanding from the SSRC's social agenda to defense-related research in the 1940s and space research from the 1960s forward. Then, in the 1970s, it was reinforced in science-based areas of high technology and subsequently aligned with health and sustainability.

All the while new fields and educational programs have continued to emerge, resulting in both Pejoration and Broadening of meaning. The principle of unity and the system of disciplinarity were increasingly challenged, and transdisciplinarity became more prominent as new connotations expanded beyond the traditional quest for unity of knowledge. In addition to general systems theory, new transdisciplinary syntheses included feminist theory, cultural critique, and sustainability as well as connotations informed by the worldview of complexity in science and by co-production of knowledge with stakeholders in society. Celia Lury also suggested social movements belong in a "third space" that includes social movements beyond the academy, the public, and representatives of government and industry (2018, p. 9) .

Even this brief opening snapshot reveals why interdisciplinarity is a conflicted discourse. Rhetorics of holism and synthesis compete with problem solving and innovation as well as critique. Opinions also differ on the foundation of interdisciplinarity, with claims ranging across epistemology and methodology to social justice and product innovation. Organizing languages have changed as well. At the first international conference on interdisciplinary research and teaching, held in France in 1970, key terms were *logic*, *cybernetics*, *structuralism*, *general systems*, and *organizational* and *information theories* (Apostel, 1972). Today the typical warrants are *complexity*, *contextualization*, and *collaboration*. Frodeman (2017, p. 3) reported the term *innovation* stood out across the 46 essays of the updated *Oxford Handbook*. Yet, even the same keyword has different connotations. Innovation, for example, is aligned with commerce, new protocols and treatments in healthcare, educational experimentation, and transformation of institutional structures and professional practices. Furthermore, commitments differ even in the same field. Weszkalnys and Born's (2013) comparison of three research initiatives provides an introductory illustration of complications for both definition and implementation.

The first initiative Weszkalnys and Barry (2013) examined, the Öko Institut, formed in the late 1970s as a scientific forum of environmental citizen groups and by the early 2000s was an independent advisory center. The second, the Tyndall Center for Climate Change Research, appeared in 2000 as a self-conscious interdisciplinary initiative across several British universities. The third, the Earth Institute at Columbia University, emerged in mid-1990s as an institutional innovation for the 21st century. The Tyndall Center sought to move beyond natural science conceptions of climate change by incorporating social aspects. However, interdisciplinarity was marginal to climate change research in the UK at the time. The Earth Institute, in turn, was conceived as a tool for problem-solving, but inter- and trans-disciplinarity were of lesser significance. In contrast, the Öko Institut identified with a problem-oriented connotation of TD beckoning a new social contract for science. It was established outside the university but faced pressure for academic performance norms such as publications. So did the Tyndall Center.

New developments also challenged interdisciplinarity as the dominant descriptor for boundary crossing. At the turn of the century, Rustum Roy's (2000) edited volume, *The Interdisciplinary Imperative,* contended that accelerated blurring of boundaries of academe, government, and industry meant "interactive research" had become a more apt descriptor. Five years later, a state-of-the-art report on interdisciplinary research from the US-based National Academies of Science, Engineering, and Medicine documented increased partnerships across the three sectors (NASEM, 2005). The following year Maton et al. proposed "intersectoral" is a more appropriate term for action-oriented research collaborations across boundaries of academic disciplines and other spheres (2006, p. 5). This book does not account for all sectors. Doing so would require a multivolume encyclopedia. Instead, it reads intersections of *interdisciplinarity* with developments and related concepts that render the core term inadequate to account for a fuller range of boundary work across disciplines, fields, occupational professions, and public and private spheres in the North and the Global South. Thus crossdisciplinary and cross-sector are key to analysis throughout the book.

Crossdisciplinary encompasses multi-, inter-, and trans-disciplinary forms of research and education. Some have stipulated specific connotations, but the term appears increasingly as a generic adjective for these and other forms of crossing disciplinary boundaries, especially in the academic sphere. *Cross-sector* refers to work bridging the multiple sectors of society identified above. The subject of definition is likely to produce groans. Rick Rylance faults "arcane debates" on terminology as "faintly theological hair-splitting," citing in particular my sorting of methodological, theoretical, instrumental, and critical forms as well bridge-building and restructuring in a typology of definitions for the *Oxford Handbook* (Rylance, 2015, p. 314; Klein, 2017). However, a baseline of multi-, inter-, and trans-disciplinarity is crucial for clarity. The Glossary includes definitions of these and other related labels, concepts, and activities. It provides snapshots of meanings that appear both within separate chapters and across them. In weighing the overarching question of definition, Frodeman (2017) suggested inter- and trans-disciplinarity are both boundary objects with different meanings at different times for different groups. Therefore, to repeat a refrain that also echoes across chapters, context matters.

The Book

This book's contention that *interdisciplinarity* is no longer an adequate term to account for heterogeneity of activities currently circulating does not amount to jettisoning the concept. Instead, it recognizes both continued use of the term and expanding interests that cross boundaries of sectors. This premise does not mean anything goes, however. It eschews essentialism of interdisciplinarity per se while concurring with the British Academy's (2016) call for defining the concept by how it is practiced. In the same vein, Jill Vickers (2003) admonished earlier challenges of navigating interdisciplinarity today require ending the search for both universal and timeless characteristics. We can better understand the concept, she urged, by studying how it is manifested in contexts. In doing so this book attempts to reorient the way we think about interdisciplinarity.

It is not the only book to try to do so. Harvey Graff (2015) claimed to offer the first critical history of interdisciplinary efforts and movements within the modern university. However, like him, others have conducted historical and comparative studies, challenged overstated positions, argued disciplinarity and interdisciplinarity are linked, accounted for institutional and organizational factors, identified conflicting definitions and purposes, criticized lack of attention to integration, and articulated the centrality of problems and questions. In the course of discussion, this book also corrects dubious claims and outright errors. Questionable claims include the belief successful initiative becomes just another discipline, an error addressed more fully in Chapter 3 on "Interdisciplinary Fields." This book also disputes the dichotomy of specialists versus generalists. Fiore, for instance, contended interdisciplinary science is essentially team science because it is not possible to conduct "truly" interdisciplinary research independently. However, he equated it with erudite thinkers such as Leonardo da Vinci (2008, p. 272). Such individuals are often dismissed as a jack-of-all-trades and master-of-none. This characterization reinscribes a false dichotomy of depth versus breadth, ignoring complexities of boundary crossing from borrowing methods and tools to hybrid specialization.

Other errors include an assertion in a state-of-the-art report *Facilitating Interdisciplinarity* that interest waned in the 1980s (NASEM, 2005). To the contrary, a more careful literature search reveals expanding interest at the time. Attacks on the core concept are also problematic. In a book marketed as a case against interdisciplinarity, Jerry Jacobs (2013) rightly countered criticism of disciplines as hermetically sealed silos by highlighting their dynamism. He was also correct to question proliferation of centers as proof of interdisciplinarity and the vague promise of application as an integrative force. However, he claimed new enclaves fail to meet the goal of bridging broad intellectual terrains. That is not a universal purpose. Moreover, not all proponents aim to overthrow the disciplinary system. In an even more full-throated attack, Graff asserted the history of interdisciplinarity is "littered with great expectations and disappointed hopes." True, but calling it "replete with sheer absurdity, wasteful competition, and hurtful personal invective" is gratuitous (2015, p. 214).

Before previewing the chapters, two explanations of methodology are in order. First, the book itself and my prior work are interdisciplinary by triangulating rhetorical, sociological, and historiographical approaches. Rhetorical analysis dissects claims by which people construct a field, patterns of consensus and difference, and keywords that shape and reinforce hierarchies of value. Sociological analysis examines how individuals and groups establish identity and control through cognitive authority and reputational systems, while historiographical analysis uncovers genealogies of origin, periodizations, and tensions between continuity and change. The three methods are not isolated, however. Genealogical studies reveal ways discursive objects, concepts, and strategies produce regularities and rules that both sustain practices and are challenged by subsequent ruptures and refigurations. And, sociological questions of power and change arise in tracking narratives of knowledge. Rhetoric looms especially large in this book because language and argument are central to definition and representation. Moreover *communication* is a keyword in the title and is crucial to the keyword *collaboration*. The prominence of rhetoric also leads to the second methodology.

Alan Liu (2008) contended interdisciplinary knowledge is itself a rhetoric that reconfigures closures in order to address new urgencies. Debra Journet (1993) also asserted true boundary rhetorics are not simply combined or juxtaposed genres. They are new genres representing new ways of thinking and acting. They do so through the first keyword in the subtitle: *boundary work*. It is associated conventionally with Thomas Gieryn's (1983) definition of the term. He treated attempts to demarcate science from non-science or pseudo-science as a rhetorical device that unfolds in three genres of boundary work: expulsion of rival authorities' claims to be scientific, expansion when rivals try to monopolize control and jurisdiction, and protection of autonomy by exploiting authority to secure resources against outside powers such as legislators and managers. Boundary construction and social identity, Gieryn added, are connected, representing shared conceptions of a group and a sense of belonging to a community.

The concept of boundary work became a standard lens for analyzing scientific groups and their rivalries, though over time its meaning also broadened because boundaries are of wide interest. Given the wide scope of contexts, this book adopts Donald Fisher's generic definition of boundary work as "a composite label for claims, activities, and institutional structures that define, maintain, break down, and reformulate boundaries between knowledge units" (1993, pp. 13–14). A few preliminary explanations about language and prior publications are also in order. To begin with, while I treat *crossdisciplinary* as a generic term for multi-, inter-, and trans-disciplinary work, when authors cite one of the terms specifically I honor their original usage though note similarities and overlaps when they appear. I also avoid the widely used terms *non-academics* and *non-scientists*. They are loaded descriptors, reinforcing status hierarchy through a negative prefix even when advocating inclusion. As for material from previous publications, I provide citations when required. However, in each case I reframe prior work around the overarching concept of boundary work while accounting for new developments and literature. Scholarship is cumulative, so updating and testing prior assumptions are crucial.

Part I: Definition

Turning to subsequent chapters, Part I of the book establishes a defining framework for thinking about boundaries, discourses of interdisciplinarity, and a major example of cross-boundary work, interdisciplinary fields.

Chapter 1: Boundary Work

Keywords: boundaries, ecology of spatializing practices, trading zones, communities of practice, boundary objects, boundary organizations, boundary agents

The first chapter lays a foundation for the book by defining boundary discourse in crossdisciplinary and cross-sector work. It begins by distinguishing spatial and organic metaphors of boundaries, with initial emphasis on disciplines. It then combines the two metaphors in a composite concept of an ecology of spatializing practices, illustrated by the evolving nature of disciplines as well as trading zones and communities of practice. The chapter then describes structures for interdisciplinary work, followed by the concept of heterarchy, the changing character of higher education, platforms for communication and collaboration, and the role of the built environment. It turns next to boundary objects, illustrated by construction of a natural history museum, an academic reform initiative, a project on waste management, and the relationship of objects and their description in climate modeling, regulatory discourse, genetic toxicology, and human ecology. The chapter ends by examining the role of boundary organizations and agents in two cross-sector case studies.

Chapter 2: Discourses of Boundary Crossing

Keywords: typology, philosophy, problem solving, critique, convergence, transformation, nomothetic, idiosyncratic

Typologies classify activities into similarities and differences in a semantic web of purposes, contexts, practices, organizational structures, and theoretical frameworks. Katri Huutoniemi and Ismael Rafols (2017) contended multiple claims tend to paralyze debate on definition. Yet, Frédéric Darbellay (2015) identified two major lines of argument in current discourse about interdisciplinarity: an epistemological, theoretical orientation that transcends disciplinary boundaries and a pragmatic, participative orientation to problem solving. The epistemic approach is philosophical, raising questions about the nature of knowledge amplified by ontological questions about the nature of reality. In contrast, problem solving is oriented to instrumental needs. This chapter compares discourses of philosophy and problem solving while adding a third imperative of critique. After acknowledging differences, it then takes into account their intersections. The chapter closes by asking whose knowledge counts, weighing the relationship of generalizations and individual cases, and reflecting on how discourse shapes definition.

Chapter 3: Interdisciplinary Fields

Keywords: field, interdiscipline, trendlines, trajectories, outcomes, change, identity, crossfertilization, intersectionality

The last chapter in Part I examines the boundary work of major communities of practice classified as fields and interdisciplines. New fields arise, Richard McKeon argued, because subject matters are not ready made to respond to all questions, problems, and issues that arise. Interdisciplinarity is thus an "architectonic art" of creating new knowledge forms and outcomes (1979, p. 18). The question of where they fit, however, inevitably arises. Lynton Caldwell (1983) argued the metaphor of fit prejudges the epistemological question at stake in formation of new fields. Many arose because of a perceived misfit of needs, experiences, information, and structures of disciplinary organization.

This chapter identifies both general patterns and contingencies of specific fields. It begins by describing catalysts, then draws insights from patterns of interdisciplinary majors in higher education and taxonomies of both research and education. It then compares trajectories and outcomes of individual cases. The following sections illustrate change over time and plural identities, then illustrate multiple themes in the chapter through the lens of women's studies and intersections across fields. It closes by asking whether there is a distinctive interdisciplinary logic.

Part II: Dynamics of Practice

Part II turns to closer analysis of dynamics of boundary work with particular attention to communicating and collaborating and learning, followed by lessons from failure and shortfalls in projects, programs, and fields.

Chapter 4: Communicating and Collaborating

Keywords: boundary weaving, translation, interlanguage, culture of communication, collective identity, integration, positioning, purposing, public engagement

This chapter explores dynamics of boundary weaving. In the process it joins Lury et al. (2018) in treating *interdisciplinarity* as a verb rather than a noun, as well as Davidson and Goldberg's (2010) recasting of institutions as mobilizing networks rather than static structures. The chapter begins by defining the nature of talk across boundaries, including pidgin and creole forms of language, linguistic and social dynamics of communication, a culture that fosters them, epistemic dimensions of dialogue, and relational thinking. It then focuses on collective identity in teams and stages of collaboration, followed by a section on integration and differing assessments of its centrality to crossdisciplinary work. The chapter turns next to public engagement and community-based research, in the process moving beyond narrow characterization of translation as application and transfer to highlight intersubjectivity, communicative action, and participatory research. It concludes by illustrating translation boundary work in two cross-sector case studies, an urban planning project and a waste management project involving both academics and community stakeholders.

Chapter 5: Learning

Keywords: integrative, constructivist, social, adaptive, generative, reflexive, transformative, transactive, competencies, experiential

This chapter presents a framework for learning across boundaries, including concepts of mutuality, interaction, and co-production. It begins with insights on integrative learning in interdisciplinary education, grounded in a shift from content-based to process-based integration, the theory of constructivism, and the concept of reflective equilibrium. After noting parallels with transdisciplinarity it examines the nature of social learning, anchoring discussion in four theoretical discourses for interdisciplinarity and interprofessionalism (Communities of Practice, Critical-Historical Activity, Complexity Science, and Actor-Network Theory). Turning more specifically to trans-disciplinary work, the chapter examines mechanisms of learning in a communication boundary space, while incorporating concepts of triple-loop learning, reflexivity, convergence, transactivity, and heuristics. Finally, after identifying individual and collective competencies, as well as characteristics of cross-sector expertise, the chapter concludes in parallel by drawing insights from case studies across sector boundaries, beginning with a sustainability project in the East India Plateau and followed by healthcare in two hospitals.

Chapter 6: Conclusion: From Failing to Succeeding

Keywords: failure, shortfalls, adhocracy, success, context, resistance, technology, evaluation, resources

Decades of reports have delineated factors for success. Yet, projects, programs, and fields continue to falter. Instead of closing the book by compiling yet another litany of recommendations, this chapter begins by condensing barriers and impediments into a digest of challenges for both crossdisciplinary and cross-sector work. It then elaborates reasons for shortfalls with examples highlighting impediments to radical forms of interdisciplinarity, questioning the litmus test of integration, and marking continuing limits to developing and sustaining fields and programs. The chapter turns next to six overarching principles for success: transparency; informed use of best practices, models, guidelines, and authoritative reports; consistency and alignment of activities in a systematic approach; balance of disciplinary, professional, crossdisciplinary, and cross-sector work; credit for boundary crossing; and appropriate criteria with a multi-methodological approach to evaluation. This section also considers the role of technology, the academic reward system, and responsibility for change. The chapter and the book conclude by returning full circle to the question of what constitutes interdisciplinarity today, followed by five gateways into the burgeoning body of resources.

PART I
DEFINITION

1
Boundary Work

> Appreciating the diverse roles that boundaries play is no easy task. It involves figuring out what boundaries enclose and what they exclude; whether they are drawn in bold, unbroken strokes or as a series of intermittent, irregular dashes.
>
> **Greenblatt and Gunn (1992, p. 4)**

> By *definition*, there is no organised and authoritative body of knowledge for academics in an interdisciplinary study, no habitual methods of judgements that have been thrashed out and debated to the point where expert discussion is even possible.
>
> **Davies (2011, p. 61)**

Boundaries pervade our lives: from geopolitical borders and legal jurisdictions to prescribed lanes in traffic and sports to divisions of academic subjects and occupational professions. Their variety is as striking as their ubiquity. They span physical and cultural parameters, social and political categories, and intellectual and pragmatic actions. Boundaries are often associated with separation and closure, but they are permeable and contingent as well. Thus, Akkerman and Bakker (2011) contended a boundary can be seen as a sociocultural difference leading to either discontinuity or interaction. A utopic strain imagines their dissolution, a hope Boix-Mansilla, Lamont, and Sato (2016) likened to the myth of the philosopher's stone that turns vulgar metals into gold. Yet, authors of the opening epigraph concluded that boundaries "can be crossed, confused, consolidated, and collapsed; they can also be revised, reconceived, redesigned, or replaced." However, they cannot be entirely abolished (Greenblatt and Gunn, 1992, p. 4).

This chapter lays a foundation for the rest of the book by defining boundary discourse in crossdisciplinary and cross-sector work. In doing so, it also counters the second epigraph. Davies (2011) was correct to admonish that definitions familiar to scholars are not necessarily read, or ever will be, by researchers and educators engaged in the work. Yet, this book is dedicated

Beyond Interdisciplinarity. Julie Thompson Klein, Oxford University Press. © Oxford University Press 2021. DOI: 10.1093/oso/9780197571149.003.0002

to informing their efforts. The chapter begins by distinguishing spatial and organic metaphors, with initial emphasis on disciplines. It then combines the two metaphors in the composite concept of an ecology of spatializing practices that acknowledges interactions between social groups and their environments, illustrated by the evolving nature of disciplines as well as enclaves of trading zones and communities of practice. The chapter then describes structures for interdisciplinary work, followed by the concept of heterarchy, the changing character of higher education, the built environment, and platforms for communication and collaboration. It turns next to boundary objects, illustrated by a natural history museum, an academic reform initiative, and a project on waste management, as well as the role of language in framing climate modeling, regulatory discourse, genetic toxicology, and human ecology. The chapter ends by examining boundary organizations and agents in a cross-sector case study of sustainable agroforestry and watershed management.

Boundaries

The word *boundaries* typically conjures up spatial images of "turf" and "territory." Along with other metaphors of location and jurisdiction, Jeffrey Peck observed, they have produced a topographical discourse that highlights "intellectual *surfaces* and academic *contours*, critical *boundaries* and scholarly *fields* of demarcated interests" (1989, p. 179). In the academic sphere the dominant bounded space is a discipline, often characterized in geopolitical terms. In an earlier discourse analysis of related images, I found talk of "private property," an "island fortress," a "fiefdom," and a "balkanized" specialty staked off by "patrolled boundaries" and "no trespassing notices." Locked in "bastions of medieval autonomy," specialists nurture "academic nationalism" reinforced by a policy of "protectionism" and a "tariff mentality" that resists "breaching" of boundaries, "alien intrusion," pilfering by "intellectual scavengers," "floundering expeditions" into other territories, and excursions to "frontiers" of knowledge (Klein, 1990, pp. 77–78). Disciplining is so deeply embedded in the academy it even determines boundaries in the same area. Teaching English, for example, would seem to be a single domain. Yet, Dressman, McCarthey, and Prior (2009) identified divisions of literary criticism, rhetoric, and writing studies; scholarship and creative writing; quantitative versus qualitative research; and education in universities and schools versus the workplace.

The different constructions Dressman, McCarthey, and Prior identified raise the question of what constitutes disciplinarity. Some trace its origin to Aristotle's distinction between theoretical and practical forms of knowledge, though in the Roman era the Latin root *discipulus* meant "pupil" and *disciplina* was associated with teaching. Others point to emergence of specialized subjects in the late Middle Ages, when the university evolved from secular cathedral schools and teaching dogmas of theology, law, and medicine. Others yet trace origin to the mid-17th and late 18th centuries, when modern sciences began assuming separate identities. By the mid-19th century, social sciences were segmenting into anthropology and economics, followed by psychology, sociology, history, and finally political science. Ironically, since philosophy is one of the oldest subjects, humanities were last to assume modern disciplinary form (Klein, 2005). By 1910, though, the university was reorganized around 20 to 25 disciplines, each exhibiting [Knorr] Cetina's (2003) concept of an epistemic culture.

Division into separate domains divided insiders from outsiders, a process Bryan Turner (1990) likened to an ecclesiastical connotation of an order maintained in a church or a medical connotation of a regimen imposed by a doctor on a patient. Messer-Davidow, Shumway, and Sylvan (1993) captured the consequences:

> Disciplines specify the objects we study (e.g. genes, deviant persons, and classic texts) and relations that obtain from them (mutation, criminality, and canonicity). They provide criteria for knowledge (e.g. truth, significance, and impact) and methods of obtaining it (quantification, interpretation, and analysis).

Disciplines also create economies of value by manufacturing discourses, providing jobs, securing funding, and generating prestige.

Raymond Miller (2020) summed up the cumulative effect. Miller described disciplines as composites of specialized subjects, techniques and methods, principles of analysis, theories and concepts, thought models, terminology, and aesthetic standards. Along with shared premises, concepts, values, and norms they produce a particular worldview upheld by a community of scholars with a shared identity, a literature, and agreement on what to teach. In the aggregate they produce a form of academic enclosure that marks boundaries upheld by agreed-upon standards. In turn, Miller added, journals, professional associations, graduate training, control of promotion and tenure, and conferral of grants anchor a hierarchical system. Stephen Turner (2017) further described a discipline as a locus of specific competencies, forms of communication, and a job market. Membership is formally sanctioned by certification and peer review reinforced in tacit modes of cognition and conduct.

Beyond these elements of disciplining, Albert, Laberge, and Hodges (2009) identified parallel conceptual frameworks. In addition to [Knorr] Cetina's (2003) epistemic culture they include Pierre Bourdieu's disciplinary habitus and Tony Becher's academic tribes. Frédéric Darbellay (2015) also cited Ludwik Fleck's concept of a thought style, akin to Thomas Kuhn's notion of a paradigm. Parallels appear in occupational professions as well, such as law, medicine, education, and social work. They include autonomy over training and certification, a designated set of knowledge and skills institutionalized in a curriculum, ethics, and a community of practice. Andreas Liljegren (2012) identified two overarching metaphors in professions. The first—landscape—highlights enclosures of social spaces, boundaries, fields, territories, turf, gates, frontiers, and maps. The second—hierarchy—treats groups as actors who maneuver to ensure positions in a pyramid such as physician and nurse, lawyer and legal aide, principal and teacher, director and employee. Comparable to disciplines, members of professions employ strategies of closure through exclusion and subordination. "Each profession," Andrew Abbott (1988, p. 33) explained in his classic work on professions, "is bound to a set of tasks by ties of jurisdiction."

Boundaries of both disciplines and occupational professions, however, are not air-tight. Introducing the second edition of *Academic Tribes and Territories*, Becher and Trowler (2001) described a number of factors driving change today: foremost among them external forces that belie the image of an ivory tower, poststructuralist theories that have challenged traditional disciplinary practices, the expanding scope of knowledge, and the cross-sector Triple Helix of academic, industrial, and governmental interactions. More broadly, Gabriele Bammer observed that disciplines "evolve, expand, merge, contract and disappear," rendering the relationship of disciplinarity and interdisciplinary research "complicated and untidy" (2012, p. 14). In the first edition of *Academic Tribes* Becher (1990) also likened disciplines to individual cells in a state of constant flux. They subdivide and recombine, changing shape and disposition. Some even exhibit an anarchic tendency to appear allied with counterparts in other domains. Yet, Stanley Fish (1989) charged in a widely read polemic that interdisciplinarity is impossible to do.

Fish acknowledged an agenda of interdisciplinarity seemed to flow naturally from left culturalist theory. Deconstruction, Marxism, feminism, new historicism, and radical neopragmatism were all critical of two kinds of boundaries: social structures that maintain political authority and institutional structures that establish and extend territorial claims of disciplines. However, he charged, any strategy that questions foundations of disciplines negates itself if it becomes institutionalized. The multitude of studies and projects center on straightforward tasks requiring information and techniques from other disciplines. Or they expand imperialistically into other territories. Or they establish a new discipline with a new breed of antidisciplinary counterprofessionals. Responding to Fish, Arabella Lyon (1992) suggested the metaphor of a river instead. Knowledge practices have currents and flows, tributaries, eddies, and confluences. Fish's position, Giles Gunn (1992) also replied, perpetuates the dualism of disciplinarity and interdisciplinarity, while Joe Moran (2002) faulted his assumptions that disciplines are coherent or homogeneous and that interdisciplinarity is synonymous with an ultimate synthesis of knowledge.

Historical perspective is also illuminating. Bender and Schorske (1997) reported many disciplines have become more porous since the 1950s. One of the major activities associated with this process is borrowing. In a genealogy of interdisciplinarity in social sciences, Craig Calhoun (2017) cited adoption of psychological testing, statistics, surveys, and network analysis across disciplines, fields, and professions. In engineering education research, Kacey Beddoes (2014) also reported incorporation of narrative analysis, case studies, grounded theory, ethnography, phenomenology, and discourse and narrative analysis. As a result, diversity gained favor over disciplinary rigor. Yet, faculty often retain traditional labels. Marcia Bundy Seabury (1999) reported many faculty who teach in the University of Hartford's interdisciplinary core curriculum remain in conventional departments. But, "[i]f you look beneath the surface you often find people who have been covert boundary crossers all along." The complexity of their interests belies the linearity of departmental affiliations (p. 5). Moreover, a sociologist studying patterns of political language in urban and rural communities would more likely read literature on rhetoric than colleagues reading about quantitative analysis of demographic databases. Yet, they often retain traditional titles as sociologists working in departments of those name.

Implications also follow for the identity of disciplines. Burggren, Chapman, Keller, Monticino, and Torday (2017, p. 101) described boundaries of biology as "constantly shifting as new technologies and theories arise, evolve, and mature and—sometimes—fade away." Biologists today are using computational algorithms engineers developed to generate predictive models of complex biological processes and systems. Collaborations with physicians and engineers have produced innovations in regenerative medicine. Genetics, molecular biology, and physiology have merged in genomics. And, biochemistry is being driven by interactions with information science and nanotechnology. The authors predicted biological sciences in the future will continue to operate in an interdisciplinary cycle, spawning new subdisciplines that change the nature of the discipline. Yet, challenges remain, not the least of which are communicating across separate disciplinary languages, constructing common knowledge, finding time for collaboration, and being evaluated by appropriate criteria. Nonetheless, the rich array of activities they reported indicate interdisciplinarity in biological sciences is literally burgeoning.

Art history is a comparable example of a decidedly different discipline. Academic study of art borrowed from other disciplines from its beginning in the 18th century. Period style was the most prominent early basis for disciplinary relations. Common motifs, themes, and genres suggested synchronic relations in chronological eras and stylistic categories such as romanticism and postmodernism. The practice dubbed interdisciplinary arts and interarts borderland also fostered hybrid forms, such as incorporating text in paintings. In the latter half of the 20th century, new critical, theoretical, and historical approaches drew on both social sciences and humanities. Interest in visual culture and digital technologies also broadened the scope of the discipline, and new hybrid genres integrated sound, image, text, and kinetic movement. Subfields such as sociology of art and feminist art history formed new interdisciplinary enclaves as well (Klein, 2005). Boundary talk in art, Ben Tilghman (2006) reflected, has typically hinged on what art is or is not, prioritizing the academic discipline over crafts, photography, popular art, and installations. An exhibit titled *Crossing Boundaries* illustrates implications of classifying. Quiltmaking, for example, is usually deemed a craft but, when moved from a decorative cover on a bed to a gallery wall, categories of art and design blur.

Increased boundary crossing is documented in updated handbooks and textbooks. Support for change, however, is uneven. In geography, for instance, B. L. Turner (2002) cited differences between a spatial-chronological structure and a human-environment conception. The former dominated until the mid-20th century, when the latter gained influence. So did earth system and sustainability science. Turner acknowledged research remains largely empirical and increasingly quantitative. However, other practices include efforts to balance agent- and structure-based frameworks, emergent properties of complexity and disequilibria in ecology and integrated assessment, as well as erosion of the boundary between pure and applied research. Even so, along with calls for reconciling spatial-chronological and human-environment positions, Turner reports the line between humanities and sciences is being reinforced. Hybrid methodologies and new barriers may also be forming around divisions of qualitative and quantitative approaches.

Disciplines, in short, are neither static nor monolithic. The relationship of disciplinarity and interdisciplinarity also exhibits both complementary and antagonistic dimensions, as well as simultaneous association and dissociation, integration and disintegration. Challenges for interdisciplinary work vary as well. Some disciplines are closer than others. Bridging biology and chemistry or history and literature, for example, is a lesser challenge than bridging disciplines with different epistemologies and methodology such as biology and literature or physics and art history. Power differentials also come into play. So-called hard disciplines, such as empirical sciences, have very different modes of operation than soft ones, such as humanities and arts though *hard* and *soft* are loaded terms, reinforcing status hierarchy. Pure or theoretical outlooks are also valued in the academy over applied ones. Proximity and distance factor into exchanges and attempts to integrate components as well. Regardless, the dichotomy of disciplinarity and interdisciplinarity has eroded.

Marcovich and Shinn (2011) even contended a New Disciplinarity is operating today, propelled by the accelerating volume and the complexity of concepts, instruments, materials, and scope of questions. They consider driving factors nothing less than the matrix of contemporary knowledge production. The trend Marcovich and Shinn labeled a New Disciplinarity is evident in novel connections and combinatorials as well as transformations of practices in older and newer domains such as solid-state and soft-matter physics, molecular biology, nanoscale research, and climatology. However, despite robust interactions they contended disciplines remain the primary referent for researchers, arguing their work at the periphery does not disrupt allegiance to a disciplinary home base and personal identity. Marcovich and Shinn also equated interdisciplinarity narrowly with only two developments—postmodernism and technoscience—and contended multi-, inter-, trans-, and post-disciplinarity all aim to abolish boundaries, ignoring recognized distinctions in their relationships to disciplines.

Other generalizations are likewise dubious. Frickel, Albert, and Prainsack (2016) described their volume of essays on interdisciplinary collaboration as a response to three beliefs underlying social scientific analysis of interdisciplinarity. They acknowledged disciplines evolve. They also worried, quite rightly, that top-down initiatives driven by administrative priorities for funding can be problematic and not even interdisciplinary. And they recognized different meanings favor particular disciplines. However, they contended erroneously that three assumptions prevail in social sciences: (1) interdisciplinarity is better than disciplinarity; (2) disciplines are silos constraining development of interdisciplinary knowledge; and (3) interactions are not constrained by status hierarchies and power asymmetries in disciplines. In a review of their book, Rick Szostak (2017) countered scholars associated with the Association for Interdisciplinary Studies and networks devoted to transdisciplinarity as well as integration and implementation sciences would not assume (1) or (3). Furthermore, in the case of

(3) scholars and practitioners have identified strategies for overcoming barriers and have taken a more nuanced approach to (2) based on evolution of disciplines and ubiquitous borrowing.

Underlying assumptions about boundaries in the foregoing examples underscore their ubiquity. In sociological research, Lamont and Molnár (2002) reported, the most general conception of a boundary is a distinction that categorizes objects, people, or activities. Boundaries, they added, have long been of interest in social theory, fostering a boundary rhetoric associated with territorialization and politicization as well as re/location, and re/institutionalization. Beyond academic social sciences, Catrin Heite (2012) noted interest in boundaries in fields as varied as science studies, gender studies, education, and social work. She also emphasized that identifying boundaries puts an analytical focus on questions about spatial borders, hierarchies and social structures, knowledge and knowledge production. In short, boundaries influence how we construct reality. Hence, she highlighted Michel Foucault's (1995) treatment of a boundary as an aspect of regimes of power.

Foucault stipulated boundaries prescribe social order, dualisms of normality and deviance, as well as belonging and not belonging. A boundary classifies, categorizes, sorts, segments, and normalizes. It also includes and excludes, privileges and de-privileges. At the same time, Heite added, no boundary remains uncrossed. Moreover, boundaries are contested areas and their authority is disputed. They are also defied, with potential to shift and transform them. Edwards and Fowler (2007) further noted increasing interest in boundaries is a result of the expansive influence of postmodernism, post-structuralism, postcolonialism, and feminism. They do not take boundaries as given, calling attention to marginal and decentered alternatives to centralized discourses of power. Hence, they align with the discourse of critique in the next chapter. C. Claire Hinrich (2008) further argued boundary work is inherent in the very concept of interdisciplinarity. Mathias Friman (2010) suggested the process of demarcating science from policy and processes, with the aim of transgressing their boundaries, has a lot in common with the debate on interdisciplinarity. And Centellas, Smardon, and Fifield (2014, p. 316) proclaimed, "The concept of boundaries is fundamental to categorizing research as interdisciplinary."

An Ecology of Spatializing Practices

The boundary rhetoric of both crossdisciplinary and cross-sector work is wide ranging. It encompasses acts of spanning, crossing, and bridging; processes of interacting, integrating, and collaborating; strategies of brokering, mediating, and negotiating; operations of demarcating, constructing, and refiguring; new relations of interdependence and convergence; and outcomes of breaching, transgressing, and transforming. Spatial metaphors alone are not enough to account for their cumulative force. Organic models compare knowledge to ecological processes. Boundary concepts, Dallimer and Strange (2015) noted, are prevalent in ecology: as edges, ecotones, boundary layers, gradients, climes, transition zones, and interfaces. In the academic world, Michael Winter (1996) explained, the parallel connotes generation, crossfertilization, interrelation, and hybridization. The English word *ecology*, he recalled, derives from a Greek term meaning "household" or "settlement." Verbs associated with *oikeos* suggest not only inhabiting and settling but also managing as well as governing and controlling. Organicism and environment, he added, might even be combined into a third model that highlights interactions between social groups and their environments. They reinforce jurisdictional claims akin to territorial claims humans and animals make in ecological niches, while creating new life forms, species, and settlements.

This chapter illustrates Winter's third type by advancing an ecological approach to crossdisciplinary and cross-sector work. An ecological approach, though, raises a topic of considerable debate, the difference between place and space. Bruce Janz (2017) cautioned that meanings vary: the terms have sometimes been used interchangeably, sometimes distinctly, and sometimes in conjunction with others, such as maps, networks, and migration. One of the most common connotations of place highlights physical location, stability, and permanency. However, Jeff Malpas (2018) advised, place is as much about connecting and being open to possibilities as it is about bounding and enclosing. Michel de Certeau's (1984) concept of space as a practiced place acknowledges both. De Certeau's focus was everyday life in a city, but his distinction between strategies and tactics is relevant to navigating knowledge. He contrasted the view of a city generated by official strategies of governments, corporations, and other institutions of power from a walker's tactics of experiencing it at street level. Tactics are influenced by formal strategies such as maps, rules, and procedures governing boundaries. Yet, walkers do not consume them passively.

Reynolds and Fitzpatrick (1999) emphasized de Certeau's concept is born of tactics of cutting across. Two sites illustrate the process: trading zones and communities of practice. Peter Galison (1997) borrowed the concept of trading zones from anthropology in order to describe how dissimilar cultures establish common ground. When bartering fish for baskets, for example, participants have different meanings of exchange. Yet, they are able to arrive at a consensus. Extending the concept to science, he cited development of particle physics and radar, facilitated by exchanges that bridged subcultures of theory and experiment. The concept of communities of practice emanated from Lave and Wenger's (1991) observations of Yucatán midwives, Liberian tailors, navy quartermasters, and meat cutters. A group with a common interest builds community by sharing information and experiences. Wenger (1998) subsequently identified three components that will figure in Chapter 4, on "Communicating and Collaborating." The first, mutual engagement, builds norms and relationships that bind individuals together. The second, joint enterprise, entails shared understanding, while the third, shared repertoire, builds communal resources.

Trading zones and communities of practice are not the only sites of cutting across, either. Accounting for current practices in interdisciplinary higher education, Karri Holley (2017) recalled in prior decades the topic occupied niche areas that may be viewed as transaction spaces, including undergraduate teaching, student learning, and curriculum development. In contrast, contemporary rhetoric positions interdisciplinarity in near revolutionary terms, aligned with transcending disciplinary knowledge and bettering the human condition. In an earlier account Klein and Newell (1997) also concluded higher education was shifting from a simple to a complex model. In the simple model, interdisciplinary (ID) work was often innovative, but its home was a familiar format or structure. A simple system might have multiple levels and connections in a hierarchy, but they operate within a single set of rules. New forms and practices were accommodated but did not challenge the existing structure. In contrast, complex systems are nonhierarchically structured, and they exhibit multiple and conflicting logics. Table 1.1 is Klein and Newell's comparison of the two models.

Activities in Klein and Newell's right-hand column, though, still intersect with the left-hand column. Heterarchy rather than homogeneity becomes a defining metaphor compatible with an ecological approach. Rosenfeld and Kessel (2008) suggested heterarchy provides a framework for collaboration in transdisciplinary problem solving, associated with mutual accountability, interdependence, power sharing, and inclusive decision-making. Relations are heterogeneous, while boundaries are fuzzy and permeable. Implications follow for how we think about institutions as well. Burton Clark's (1995) comparative study of research universities revealed they were grappling with a gap between older, simple expectations and new developments that outrun them. He concluded definitions depicting one part or function of a university as its essence or essential mission obscure changes that are transforming knowledge and education.

Table 1.1 Visible and invisible forms of interdisciplinarity on higher education campuses

Visible Forms	Invisible Forms
• Freestanding institutions	• New ID approaches and trading zones
• Autonomous colleges within universities	• Changing disciplinary curricula
• Centers and institutes	• Learning communities and faculty networks
• Departments and programs	• Problem-focused research topics, projects
• Majors, minors, and concentrations	• Shared facilities, databases, instruments
• Self-designed majors	• Subdisciplinary boundary crossing
• Mainstream and alternative general education	• Educational functions of centers
• Individual course within disciplines	• Training for ID and collaborative modes
• Tutorials and independent study	• Collaborative research programs, projects
• Travel-study, internships, and practicums	• Intersectoral partnerships with government, industry, and communities

Source: Klein and Newell, 1997, pp. 397–399; Laursen Evaluation and Design, LLC, 2021.

Trowler and Knight's (2002) studies of institutional change shed further light on the gap Clark identified (based on Klein, 2010). The standard model of "contextual simplification," they reported, assumes organizations are culturally simple, fitting into pigeonholes. Yet, institutional cultures are protean. "[A]ny university," Trowler and Knight observed, "possesses a unique, multi-layered, and dynamic cultural configuration which renders depiction difficult and simple depictions erroneous." Viewed from an analytical telescope, differences look small. Yet, viewed from a microscope they loom large (p. 143). Given the growing complexity of organizations, proposals for restructuring the academic system inevitably emerged. In 1972 Erich Jantsch envisioned redesigning the university around systems-design laboratories, function-oriented departments, and discipline-oriented departments. Two decades later Robert Costanza (1990) imagined permanent colleges, departments, or programs of integrated, trans-disciplinary studies. And, in 2008 the *Handbook of Transdisciplinary Research* documented formats ranging from small research projects and educational programs to national initiatives (Hirsch Hadorn et al., 2008). One particular model has attracted wide attention: Arizona State University's (ASU) federated design of trans-disciplinary departments, centers, institutes, schools, and colleges.

Even a highly touted model does not transform a system of higher education. However, it models strategies for reform. ASU unbundled and dispersed disciplines across units. Biology, microbiology, and plant biology departments, for example, were merged with a program in molecular and cellular biology to form a School of Life Sciences. The new structure is organized into seven faculty groups spanning biomedicine and biotechnology; cellular and molecular biosciences; genomics, evolution, and bioinformatics; ecology, evolution, and environmental science; human dimensions of biology; organismal, integrative, and systems biology; and basic medical sciences. More than 100 life scientists, engineers, philosophers, social scientists, and ethicists work in this transaction space. Even so, ASU president Michael Crow and co-author William Debars (2017) cautioned, academic culture at large is still dominated by disciplines. The result is limited capacity for responding to complex environmental, social, and health problems. The Department of Thematic Studies (TEMA) at Linköping University in Sweden is another celebrated exemplar.

TEMA was inspired by student movements in the 1970s and undergraduate interdisciplinary degrees at Roskilde University in Denmark. Faculty at Linköping created a doctoral program instead. However, Lykke and Last (2018) recalled, establishing the unit did not happen without political struggle. Traditional units still exist, but, they concluded, TEMA has made the campus more interdisciplinary-friendly. It now houses gender as well as environmental/science and technology/and child studies. Beyond Arizona State, TEMA, and other exemplars, two additional strategies facilitate cutting across boundaries. The first—constructing new and retrofitting existing buildings—colocates individuals and teams, projects, and core facilities, increasing the likelihood of crossing paths with others from different disciplines and fields. Holley (2017) reported an increase in the number of buildings devoted to interdisciplinarity over the past decade, especially in science, technology, engineering, and mathematics (STEM). Among examples she cited Stanford University's Yang and Yamazaki Environment and Energy Building and Pennsylvania State University's Life Sciences Building. The 2013 master plan for Antioch College included a science building renovated to make space for arts.

Major reforms are rare. New structures more often are circumscribed by the strength of entrenched boundaries. Top-down administrative priorities tied to grand challenges of the day have led to high-level mandates for crossdisciplinary and cross-sector work. Russell, Wickson, and Carew cautioned they run the risk of establishing new "mega-silos" (2008, p. 465). Yet they constitute a space of transactions that is not only physical. Edward Soja (1989) characterized space as a dynamic product of translations, transformations, and experience. Carla Schatz, director of the BioX Institute at Stanford University, also emphasized a common location is both a gathering point and a recruiting tool (cited in NASEM, 2005). That description applies to the second strategy as well—platforming. Hindenlang, Heeb, and Roux (2008) defined platforms as loosely structured social networks for communication, mutual understanding, shared goals, and relational capital. Comparably, Boix-Mansilla, Lamont, and Sato (2016) argued that sociocognitive platforms for interdisciplinary collaboration are shared spaces for defining problems and exchanging expertise, generating shared language, concepts, and interactions that foster a group culture and identity. They also facilitate comfort with unknown terrains and help to recalibrate beliefs about other disciplines.

Boundary Objects and Boundary Discourse

Taken together, concepts of spatializing, trading, building community, and platforming raise a question that resonates throughout the remainder of this book: How do people actually work together across boundaries? One of the most prominent mechanisms is boundary objects. In contrast to Gieryn's original definition of boundary work, Hauke Riesch (2010) pointed out, Star and Griesmer's (1989) concept of boundary objects did not involve rival groups. They proposed the concept in order to account for the role artifacts play in enabling communication and cooperation between groups despite differing values, norms, and aims. They also endorsed ecological analysis to account for the variety of institutions and viewpoints of the heterogeneity of actors who are involved.

Four types of objects help mediate across them:

Repositories: compilations of objects such as a library or a museum collection indexed in standardized fashion for use or borrowing as needed for different purposes;

Ideal Types: a diagram, an atlas, or other descriptions abstracted from separate domains but adaptable to local sites, so providing a road map for participants;

Coincident Boundaries: common objects such as creation of a geographical map by different experts even when internal contents may differ;

Standardized Forms: indices such as a map or a graph, ideas, and concepts that facilitate communication across work groups.

As the four types suggest, objects may be concrete or abstract, visual or cerebral, and material or processual. Star and Griesmer's exemplar of constructing Berkeley's Museum of Vertebrate Zoology involved amateur naturalists, professional biologists, the general public, philanthropists, conservationists, academic administrators, and taxidermists with a shared interest in California wildlife, terrain and physical environment, habitats of species, and conservation. Translation played a role in two ways: (1) developing, teaching, and enforcing methods to discipline information multiple participants collected and (2) generating boundary objects to maximize communication between social worlds. In proposing the concept, Star and Griesmer adapted Latour, Callon, and Law's model of *interessement*, meaning "interposition." Many trappers were *interessed* by amateur collections of museum workers with little to no interest in conservation or science. Yet, standardized protocols channeled collective attention to habitat, time, and geographical locations.

Every initiative, however, has not been successful. Magnus Hultén's (2013) case study of pedagogic discourse underscores Eden, Donaldson, and Walker's (2006, p. 1063) contention that a boundary is a "fuzzy zone of negotiation and rhetoric." Introduced in 1969 as part of a new national curriculum, the boundary object Spheres of Work was a thematic approach to syllabus content for one or more subjects in grades 7 through 9 in Sweden. Comparable to other cases, individuals and groups interpreted the concept differently. Yet, Hultén recalled, it was an active concept at the organizational level and a means of steering past disagreements on subject-based versus an integrated approach and student-centered interests. Groups working on social studies and science, in particular, developed an elaborate structure that moved from a single-subject approach to Spheres of Work units integrating either science or social studies to combine their content. Survey data, though, revealed differences by grade, teachers' experience, and location. Moreover, syllabi and schedules were typically structured around subjects, so teachers had to construct integrated units themselves. Ultimately, outcomes did not match hopes. In 1980 a new national curriculum replaced Spheres of Work with another thematic idea. Yet, Hultén rejected the dichotomy of success versus failure as an overly simplistic measure of a boundary object.

Spheres of Work failed to achieve widespread implementation, but Hultén concluded it had a major impact on the structure of knowledge in the 1969 reform, textbooks written in the 1970s, and classroom practice. It pushed educational discourse in Sweden further in the direction of integration, while uniting stakeholders with different allegiances and bringing scattered pedagogical discourse into a common focus despite continued favoring of progression in a single subject, especially in science. And it introduced aspects of progressive education into subject-based teaching. This case is not isolated, Hultén added. The project method, groupwork, collaborative learning, and networked learning have all enjoyed support, and the STEM movement is currently promoting integrated approaches. However, goals for STEM differ. Some contend it will increase student learning and employability, while others critique its social and economic rationales. Others yet cite lack of integration across subjects, while raising concerns about marginalizing humanities, arts, and even social sciences.

Pohl, Fam, Hoffmann, and Mitchell (2019) further illustrated boundary objects in a water management project aimed at creating a new sanitation system in a building at the University of Technology Sydney. Urine diversion systems separate and collect urine from wastewater streams for processing as fertilizer. Six disciplinary faculties, five industrial sectors, and three governmental departments had to learn about related technological, social, and regulatory factors. Two boundary objects fostered transdisciplinary and action research: a schematic of the conceptual approach and a systems diagram. The first object mapped five strands of potential inquiry, while the second helped identify everyone's contributions and connections. The project was not limited, either, by a linear workflow with designated milestones and predetermined outcomes. It employed a horizontal structure of leadership, recognizing experience and expertise of all participants. Continual monitoring was aided by weekly reflection on the project's development by core members, monthly meetings in each research strand, and meetings of the entire group every four to six months. Even so, after two years outcomes were primarily academic conference papers, not long-term institutional change.

Beyond illustrating both the value and the limits of boundary objects, they also affirm the centrality of language. In a case study of genetic toxicology, Scott Frickel (2004) recalled how a group of geneticists reordered conceptual and social contours of genetics by reframing the boundary object of synthetic chemicals in human environments as genetic hazards. Before 1966, genetic toxicology and environmental mutagenesis were not part of life science discourse. Through amplification, though, the geneticists clarified a new interpretive frame, and through extension encompassed interests or points of view once deemed incidental. Frickel also invoked Bruno Latour's concept of translating interests to connect differing points or highlight areas of similarity and association, thereby mobilizing interactions in laboratory and regulatory practices. In the case of genetic toxicology, creating a new interdiscipline entailed politicizing the meaning of chemical mutagens to prompt action by biologists. Creating a framework for processes of social organization and change also appropriated ideas from biology in order to confer scientific identity on human ecologists. In a comparable example, Emanuel Gaziano (1996) recalled sociologist Robert E. Park's effort to locate sociology as a science related but not subordinate to biology. He did so by differentiating the discipline from social work and the interdisciplinary field of urban studies while helping position the University of Chicago's School of Social Science.

The entwined boundary work of rhetorical positioning was further apparent in emergence of nanotechnology. Kaplan and Radin (2011) called determining its meaning a political process of negotiation with implications for practice, funding, legislation, and stakeholder support. Here, as in other examples, emerging schools also constructed the field differently. K. Eric Drexler's argument for an engineering-based research program was based on developing molecular manufacturing, while Richard Smalley favored a chemistry-based approach. Echoing Kaplan and Radin, claimants to all fields produce texts that anchor discourses, shaping how it develops and establishes boundaries of legitimacy and validity that distinguish what is in and what is out. Two additional examples further underscore the centrality of discourse in defining a field.

The first example is the emergent discipline of educational neuroscience, which aims to integrate disciplines investigating human learning and development. Catherine and Miriam Beauchamp (2013) reported recurring vocabulary revealed boundary principles shaping educational neuroscience discourse, including connections and bridges between disciplines, inter-, multi-, and trans-disciplinarity, and tools such as boundary objects. Challenges appear, however. Learning, for example, does not have the same meaning for a neuroscientist and an educator. Also, knowledge is viewed as a received phenomenon in science versus a constructed phenomenon in education. In addition, epistemics conflict. Education leans toward rationalism and holistic and subject perspectives, while science favors analytical and objective empiricism. The Beauchamps suggested, though, educators could formulate research questions that improve application of neuroscientific findings to classrooms. For their part neuroscientists could develop greater awareness of educators' practical concerns, rendering conclusions more targeted and realistic. At present, linking is happening in the Mind, Brain, and Education and other movements that foster exchange.

The second and extended example is Kuntz and Petrovic's (2011) account of cognitive framing in teacher education. The field has changed historically from discipline-specific studies to interdisciplinary social foundations. Yet, they reported, instruction in content of subject matters and technocratic standards are pitted against expertise in pedagogy deemed pejoratively to be the work of generalists. As a result, social, historical, and philosophical foundations are increasingly marginalized in colleges of education, eclipsed by neoliberal values of efficiency, accountability, practicality, productivity, and control. Kuntz and Petrovic identified four strategies for survival: building/crossing disciplinary borders and reframing foundations; replacing traditional foundations with something else, such as multiculturalism; reframing teacher education to make its raison d'être central and reframing standards to be consistent with or even aid placement of foundations; and reasserting borders.

Strategies of crossing, replacing, reframing, and reasserting will be familiar to members of other fields as well. So will the three larger categories of communication, visibility, and practicality that subsume them. Here too, naming plays a role. One faculty member Kuntz and Petrovic interviewed reported changing a course title, "The School in United States Society," to "Critical and Contemporary Issues in Education." They likened renaming to smuggling marginalized practices into institutional structures in order to protect them, including using neoliberal tactics such as anchoring in empirical research and well-developed standards. Visibility reasserts professional authority or, at least, relevance and in this particular case a foundations form of interdisciplinarity. Related activities include serving on committees, creating spaces within and beyond the curriculum, and organizing events, groups, and discussion forums. Practicality, in turn, involves translating jargon and clarifying ideas for classroom use. Reflecting on the case study, Kuntz and Petrovic identified two types of strategic boundary work—integration and differentiation. They personally favored a form of disciplinarity that would increase potential for foundations to disrupt technocratic, neoliberal forms of teacher education. However, they did not advocate return to the simplistic notion of a gilded age, rather situating it within an expanded form of interdisciplinarity heeding socioacademic contexts.

Boundary Organizations and Agents

In addition to boundary objects and rhetorical positioning, boundary organizations and agents play an important role. David Guston (2001) defined boundary organizations as entities with interests in both science and policy. People also perform a brokering role. Long, Cunningham, and Braithwaite (2013) called them intermediaries who reach across structural holes or gaps. In a literature review of articles spanning a wide range of settings, the authors identified four cross-secting themes: identifying brokers and brokerage opportunities, generating and integrating innovation, brokering knowledge, and gaining trust. Brokers facilitate transactions and flow of information between people or groups lacking either access to each another, or alternatively, any basis for trust. Social network analysis techniques help identify the structure of the network, yielding a picture of communication flows as well as bottlenecks. In a study of leadership for trans-disciplinary teamwork

Barbara Gray (2008) also called brokers conflict resolvers who cross boundaries of mismatched knowledge, paradigms, and experience. Organizations and agents may operate in tandem as well. Case studies affirm these general observations.

Lang et al. (2012) recalled a project on conservation of Arabica coffee in mountain rainforests of Ethiopia. It was initiated by an Ethiopian scientist who conducted research on rainforest loss in his home region with the aim of creating applicable knowledge. The nonprofit Ethiopian Coffee Forest Forum, founded during the first phase of the project, though, also functioned as an intermediary between research and implementation. McNie et al.'s (2008) report on sustainable agroforestry and watershed management in Indonesia provides an extended composite example of organizations, agents, and objects. Participants reflected on boundary work in a 2007 meeting of a team from the Sustainability Science Program at Harvard University and a World Agroforestry Center (ICRAF) program on Rewarding Upland Poor for Environmental Services (RUPES). Watersheds embody both literal and figurative boundaries, determining the flow of water while also marking stakeholders' interests. Yet, boundaries blur.

Blurring occurs in both the North and West and the Global South, but in Indonesia, McNie and colleagues reported, many forestry and agricultural scientists working in national ministries are constrained by political forces that predefine problems and research agendas. Their experience in the developing world also indicated individuals often serve the role organizations perform in the North and West. An organization, though, can still transfer legitimacy and credibility. As an international body, RUPES is considered apart from local, regional, and national politics. ICRAF could also leverage its social capital if conflict arose with RUPES. For their part, boundary agents often enjoy social capital because they live in a community. Regional forestry extension agents interested in facilitating development of a community forest agreement, for example, already enjoyed the trust of coffee farmers but lacked intellectual, electronic, and financial capital. So ICRAF provided training in GPS and GIS technologies, knowledge of agricultural science, computer equipment, and money for transportation. In another case, ICRAF lacked agents with established social capital, so college graduates with training in forestry, agriculture, and development moved into communities. Five ICRAF employees and NGO partners resided in a village five days a week over a period of six months. In the process, they gained farmers' trust, enhanced by speaking the same language as the farmers because they were members of the same ethnic group.

Boundary objects also played a role. The concept of integrated water resource management, for example, facilitated resolution of conflicts around land use. RUPES' hand-drawn cartoons of communities were simple objects, but they facilitated residents' description of their concerns. Maps using GPS and GIS technologies also yielded simple but exact measurements and representations of coffee gardens in the Sumberjaya district of West Lampung in Sumatra. In contrast, a community forest agreement was a complex boundary object that required negotiating the meaning of trees. The government was concerned about farmers' use of protected forestland, maintenance of watersheds, and quality and quantity of upstream water for downstream users. Officials believed they and their coffee gardens harmed watersheds through deforestation and poor agricultural practices.

Negotiation was exacerbated by the history of relations between the government and farmers, including soldiers' eviction in the mid-1990s of thousands of farmers. However, each of five ICRAF employees and NGO partners resided in a village for five days a week over six months. The key boundary question was what type and amount of trees to plant to achieve governmental protection without compromising farmers' ability to make a living. Farmers believed planting about 1,000 timber trees/ha would crowd their gardens and create too much shade. They also wanted to plant more fruit and nut trees to enhance income. Arriving at a decision required bridging academic, political, and Indigenous knowledge. Academic science-based research determined that coffee trees are beneficial for stabilizing and protecting watersheds. In addition many fruit tree species can provide watershed protection. For its part, ICRAF demonstrated shade is not inherently bad for coffee. Government officials came to view farmers as partners rather than obstacles in protecting the watershed by helping improve conservation, preventing illegal logging, and fighting forest fires. Ultimately, resolution of differences in the community forest agreement required farmers to plant 400 trees/ha and allowed up to 40% of fruit and nut trees.

Several lessons emerge from this and other cases. To begin with, increased focus on interactions opens up a processual view of knowledge production and social relations, rather than a narrow focus on structures, products, and rule-bound behaviors. Akin to the duality of boundary objects, participants still make jurisdictional claims, but exchange occurs in trading zones and transaction spaces. Clark et al.'s (2011) analysis of natural resource management programs furnishes added lessons conducted under the auspices of the Consultative Group on International Agricultural Research's alternatives to the slash-and-burn program. The simplest boundary work they identified occurs when the purpose is enlightenment. It requires constructing shared perception on all sides of a boundary regarding credibility of the knowledge being produced. Thus, it is akin to Gieryn's demarcation of science from nonscience that is reinforced in peer review. The second type occurs when making a decision for an individual user. Salience is an added factor when interactions occur between scientists and farmers and between scientists and policymakers. When the purpose is negotiation, boundary work widens to include not only salience and credibility but also legitimacy for parties with different political interests. This study also revealed differing approaches to accountability. In some instances it is formally codified but in others, especially in the developing world, more informal and fluid.

To recapitulate, movement is as important as structure, a view of interdisciplinarity that lies at the heart of Lury et al.'s (2018) *Routledge Handbook of Interdisciplinary Research Methods*. This mammoth tome embraces the gerund. Treating interdisciplinarity as a verb rather than a noun highlights the "do-*ing*" of methods. It also parallels Davidson and Goldberg's (2010) recasting of institutions as mobilizing networks' static structures. And it underscores *Handbook* contributors Fensham and Heller-Nicholas' (2018) characterization of interdisciplinary research processes as a space of ideation akin to a limen, a threshold, or an interstice as well as nodes in a network and space often characterized as "inter-" or "in between." Interdisciplinarity is thereby situated in a "spatiality of communication" (pp. 31–32). Echoing Catherine Lyall (2019), it is time to stop seeing interdisciplinarity as an epiphenomenon to disciplinary excellence, shaped by contingencies and contexts in all sectors.

Barry and Born (2013) further cautioned against contrasting homogeneity and closure with heterogeneity and openness. Disciplines are also heterogeneous and open, at the same time interdisciplinary initiatives can result in closure. The remainder of Part I of this book will funnel lessons about boundaries from this introduction to boundary work. Chapter 2 identifies three major patterns of argument marked by their intellectual and organizational parameters. Chapter 3 then examines the complexity of establishing new communities of practice. In each case bounding and institutionalizing are prominent topics. Reviewing literature on boundaries, coupled with insights from a case study of the fishing industry, Zietsma and Lawrence (2010) identified four cycles of stability or change:

1. Institutional stability, involving boundary and practice maintenance;
2. Institutional conflict, involving breaching and bolstering the boundary and disrupting and defending practices;

3. Institutional innovation, involving establishing experimental boundaries that were protected from institutional discipline and inventing new practices;
4. Institutional restabilization, involving crossboundary connecting and practice diffusion.

Previous studies, Zietsma and Lawrence noted, been hampered by a tendency to think in terms of either change or stability. Yet, boundary work and practice work are interwoven. They operate in recursive configurations that underpin cycles of stability and challenges to the status quo. If social actors are dissatisfied with existing practices, they may engage in work to affect them directly. However, if boundaries prevent such action, they might first engage in boundary work to create conditions under which they can influence practice. In both cases, they navigate and negotiate transaction spaces through argument (Chapter 2) and field building (Chapter 3). Hence, they are caught up in a process of both bounding, unbounding, and rebounding. Following Zietsma and Lawrence, whether institutional arrangements endure or are transformed depends on the status of boundaries and practices and the existence of actors able to leverage circumstances.

2
Discourses of Boundary Crossing

> The endless typologies, classifications, and hierarchies of multi-, inter-, and transdisciplinarities are not helpful.
>
> **Harvey Graff (2005)**

> [D]iversity should not be interpreted as disagreement.
>
> **Henrik Von Wehrden, et al. (2018)**

Typology brings order to entities, whether sorting artistic genres, medical symptoms, animal species, or domains of knowledge. The prevailing system of classifying knowledge in the West for much of the 20th century was based on disciplines. In the closing decades, though, it was supplemented and challenged by crossdisciplinary activities. The opening epigraphs pit two views of definition against each other. In the first Harvey Graff blames the "name game" for generating more confusion than clarity. The second is from a report of a workshop on definition sponsored by the INTREPID network of researchers, practitioners, and policymakers from 27 countries. Von Wehrden and colleagues found overlapping words for crossdisciplinary approaches, including *cooperation*, *trust*, and *integration* as well as *system*. Even with overlaps, though, interdisciplinarity was often associated with connectedness and combination, and accentuated exchange, integration, and mutual understanding. Transdisciplinarity, in contrast, was most often associated with normative connotations, signaled by keywords of *sharing*, *co-creating*, and *co-producing knowledge*. The authors cautioned, however, that any general definition is complicated by involvement of actors outside the academy.

To reiterate, Huutoniemi and Rafols (2017) contended multiple claims tend to paralyze debate on definition. However, discourse analysis reveals clear patterns of consensus in networks and communities of practice. Discourse analysis is the study of written, spoken, and other forms of communication. Tuen A. van Dijk (2011) tracked the field of discourse studies to classical rhetoric though attributed its modern form to anthropological, sociological, linguistic, and psychological interests from humanities and social sciences

Beyond Interdisciplinarity. Julie Thompson Klein, Oxford University Press. © Oxford University Press 2021.
DOI: 10.1093/oso/9780197571149.003.0003

during the 1960s and 1970s. As the field evolved it came to encompass analysis of how media construct meaning, including new forums and modes of writing and speaking on the internet. Hence discourse analysis spans not only linguistic but also social, psychological, political, cultural, and technological dimensions of communication, further examined in Chapter 4, on "Communicating and Collaborating." The level of focus varies, though: from words, phrases, and sentences to texts and speeches to patterns of theory and practice that constitute a discipline or field. This book concentrates on the latter two levels: use of specific terminology in texts and broader similarities and differences in theory and practice.

Frédéric Darbellay (2015) identified two major lines of argument in current discourse of interdisciplinarity: an epistemological and theoretical orientation that transcends disciplinary boundaries and a pragmatic and participative orientation to problem solving. This chapter compares discourses of philosophy and problem solving while adding a third imperative of critique. It next considers their intersections, then closes by asking whose knowledge counts and weighing the relationship of nomothetic generalizations and idiosyncratic cases. In the course of discussion the chapter also recognizes a number of trends that have implications for definition of interdisciplinarity. Recalling the Introduction to this book, they include linguistic Amelioration in the heightened priority of problem solving over epistemology, and Broadening in the ascendancy of transdisciplinarity and increased alignment with crossing sector boundaries. Narrowing is also evident in the shift from a theoretically-based approach and hierarchical classifications to empirical and descriptive cases.

Philosophy

Epistemology has been the dominant focus in philosophical discourse on interdisciplinarity, traced in the West to ancient Greece and the idea of unity. This idea persisted in efforts ranging from the medieval Christian summa, Leibniz's *mathesis universalis*, and the Enlightenment ambition of universal reason to Alexander von Humboldt's holistic view of education, the Unity of Science movement to integrate scientific statements into a common foundation and terminology, the search for unification theories in physics, and individual propositions such as Umberto Eco's speculation on a perfect language and E. O. Wilson's theory of consilience. Against this backdrop, a pattern of argument developed in the history of Western thought extending from Plato to Hegel. Founded on principles of absolute truth, duality, and determinism, James Welch (2011) recounted, it paved the way for emergence of interdisciplinarity. From the beginning, the concept called into question the traditional binary structure of dualism, prescriptive determinism, and an eternal, immutable standard. The polyvalence of concepts, Welch added, is a defining characteristic of interdisciplinarity today that challenges a monological Cartesian perspective.

Darbellay (2019) further explained both interrogation of past constructs and current focus on complexity are positioned epistemologically against atomization as well as unification of knowledge. The ascendancy of transdisciplinarity (TD) has implications for philosophical discourse as well. The term is dated conventionally to the first international seminar on interdisciplinarity. The core definition was "a common system of axioms for a set of disciplines" that transcends the narrow scope of disciplinary worldviews, exemplified by anthropology as a broad science (Apostel, 1972, p. 26). Seminar participants differed, however. Andre Lichnerowicz (1972) advocated the Mathematic as a universal interlanguage. Jean Piaget (1972) treated TD as a higher stage in the epistemology of interdisciplinary relations based on a general theory of systems or structures, while Erich Jantsch (1972) imbued it with social purpose in a model of science and education with feedback across systems labs and function- and discipline-oriented departments. Piaget considered transdisciplinarity still a dream, and Jantsch did not find institutionalized examples of a purpose-oriented model at the time. However, Jantsch cited a form of normative interdisciplinarity in cultural and environmental sciences, human biology, community sciences, creative communities, and engineering and applied sciences.

Reflecting on the history of discourse on transdisciplinarity, philosopher Joseph Kockelmans (1979a, 1979b) reported it has tended to center on educational and philosophical dimensions of sciences. Philosophers differ, though, on whether the focus should be an all-encompassing theory or a unified worldview. They also differ on whether the weight should fall on philosophical reflection, critique, social relevance, or the role of science in society. All conceptual frameworks associated with the concept, Raymond Miller (1982) further observed in a typology of interdisciplinary approaches, claim to transcend the narrow scope of disciplinary worldviews. However, proponents differ on whether their frameworks should replace existing disciplinary approaches or offer alternatives. In addition, they claim different types of isomorphism with the real world they purportedly represent and greater or lesser receptivity to quantitative manipulation and empirical application.

By the final decade of the 20th century, the meaning of transdisciplinarity was also undergoing Broadening. It became associated with synthetic paradigms, including general systems theory, neo-Marxism, feminist theory, post/structuralism, cultural critique, and sustainability. It was a label for broad fields and synoptic disciplines, including philosophy, history, anthropology, and geography. And, it had become a descriptor for a team-based holistic approach to healthcare as well as integrated curriculum design. Two new definitions also arose. The first is Basarab Nicolescu's (2002, 2010) theory informed by the worldview of complexity in science. Nicolescu acknowledged difference in calling definition a war. The axiomatic methodology of this particular definition consists of multiple levels of reality, the logic of the included middle, and complexity. Plurality and unity co-exist, and every level is characterized by incompleteness. Hence, knowledge is always open. From an etymological perspective, Nicolescu added, Piaget retained connotations of being *across* and *between* in the Latin prefix *trans-*. He proposed, instead, including *beyond* disciplines, thereby including the transcultural and transreligious nature of TD in a new form of spirituality.

The second definition arose in environmental research across German-speaking countries during the late 1980s and early 1990s. It embraced Jürgen Mittelstrauss' (1992) mandate to frame research around problems of the life-world rather than disciplines. Its traits, Moritz Engbers (2018) explained, include societally relevant problems, a heterogeneity of social actors and strengthened relationships, parallel heterogeneity of their knowledges, learning as part of research, and co-production of knowledge with stakeholders in society. As the variety of examples suggests, warrants for both inter- and trans-disciplinarity continue to differ. The Institute for Interdisciplinary Research, for example, aims to integrate disciplines in traditions of liberal arts education, unity of Renaissance learning, and transcendental values of Christianity (https://www.jis3.org). The same year Nicolescu's Charter of Transdisciplinarity was signed, though, Mode 2 theory advocated socially robust knowledge, not scientifically reliable knowledge alone.

Despite continuing dominance of epistemology within philosophical discourse, Barry and Born (2013) elevated ontology to one of three defining logics of interdisciplinarity. It sheds light on how knowledge practices intervene in the world, they explained, while also generating hybrid or relational objects and entailing social arrangements when recognizing perspectives of stakeholders beyond the academy. An ontological approach was evident in a project on flood-risk management in the UK that also illustrates challenges to achieving transformative structures and practices many initiatives have experienced even when receiving high-level endorsements and funding. Recalling the project, Sarah Whatmore (2013) classified it as a form of *Radical Interdisciplinarity*, in contrast to *Superficial* or *Functional* forms. It aimed to foster integrative interdisciplinary research, to build capacity, and to transfer knowledge of stakeholders. Tensions persisted, however, between ontological imperatives, which are less developed in environmental research, and normative goals, which sidestep philosophical questions about the relationship of natural and social worlds. An ontological division of natural and social sciences also has significant implications for which practices prevail without reflection on underlying assumptions. As a result of limits, Whatmore reflected, they "become dulled in the hardwiring" of flood management policy and practice (p. 170).

Discussion of philosophical stances would be incomplete without acknowledging individuals who are dubbed inter- or trans-disciplinary in the vein of Leonardo da Vinci, who symbolizes a synoptic thinker capable of spanning the breadth of knowledge. However, the amount of knowledge to comprehend was smaller in Leonardo's time. Furthermore, many well-known synoptic thinkers have focused on clusters of disciplines and paradigms without positing a new universal theory of knowledge. Leading examples include Michael Polanyi, Wendell Berry, Michel Foucault, Stephen Jay Gould, and Isaiah Berlin. Social theorists have also constructed overarching frameworks to bring unifying perspectives to social sciences, including Émile Durkheim, Georg Simmel, Max Weber, Robert Park, and Talcott Parsons. Camic and Joas (2004) also cited thinkers who posited new theoretical frameworks: including Anthony Giddens' effort to create a synthesis of structuration theory, Randall Collins' call for a comprehensive theory of society, Jeffrey Alexander's effort to converge sociological theories, Alain Touraine's call for reunification based on a general representation of society and vision of change, Jürgen Habermas' theory of communicative action, and Niklas Luhmann's synthetic framework for analysis of social systems.

Problem Solving

A 2005 state-of-the art report on interdisciplinary research cited four major drivers: the inherent complexity of nature and society, issues not confined to a single discipline, generative technologies, and the need to solve societal problems (NASEM, 2005). The last driver has amplified the discourse of problem solving. However, it has a long history traced conventionally to the Social Science Research Council's work on societal problems in the 1920s and followed by defense-related research in the 1940s, in the latter case including the Manhattan Project to build an atomic bomb. The emerging field of operations research also became an exemplar of both crossdisciplinary and cross-sector problem solving. In 1944, Brozek and Keys reported the rise of cooperative research in science, and in 1975 Lawrence Bass recalled interdisciplinary task force management had long been a feature of civilian affairs, engineering projects, feasibility studies, and industrial research and development. From the late 1970s forward, though, the discourse of problem solving heightened in science-based areas of international economic competition such as engineering, manufacturing, computers, and biomedicine.

Given this history, to echo Steve Fuller (2017), interdisciplinarity is linked in no small part to war and commerce. Both prioritize use-inspired research, making the Triple Helix model of state-industry-university relations a benchmark in the history of the word. In the closing decades of the 20th century, though, both inter- and trans-disciplinarity became aligned with sustainability and health, reinforcing the priority of problem solving while illustrating linguistic concepts of Amelioration and Broadening. By 1982 authors of a report on changing relationships of the university and the community declared Exogenous Interdisciplinarity had gained priority over Endogenous University Interdisciplinarity. The Endogenous originates within science and the academy, with the aim of unity. In contrast, the Exogenous originates in pragmatic problems of communities and the demand universities perform a social mission (Center for Educational Research and Innovation, 1982). To no surprise, then, Peter Weingart (2010) contended research is no longer a search for basic laws but occurs in contexts of application and use. As a result, criteria and quality are no longer determined solely by disciplines, and knowledge production is more socially accountable. Amelioration and Broadening are further linked with professionalizing in a number of organizations.

The organization known as ATLAS, for example, is committed to social, environmental, economical, and ethical sustainable development, achieved by integrating knowledge and skills from all disciplines in complex systems and engineering. ATLAS is the acronym of the Academy of Transdisciplinary Learning and Advanced Study. Founded in 2000, it shares an approach to change that does not beckon radical transformation of the university. The organization contends it is not necessary to reorganize entirely. Instead, alternative structures are needed to foster collaborations (theatlas.org). The Network for Transdisciplinary Research has also built on the concept of transdisciplinary problem solving promoted at a groundbreaking international conference in 2000 that Jahn, Bergmann, and Keil called a "normative turn" in prioritizing real-world problems (2012, p. 2). Case studies were reported in all fields of human interaction with natural systems (e.g., agriculture, forestry, industry, megacities) and technical development (e.g., nuclear and biotechnology, genetics) as well as fields where social, technical, and economic development interact with value and culture (e.g., aging, energy, healthcare, nutrition, sustainable development, landscape, housing and architecture, urban land, waste management) (Klein et al., 2001). Kessel and Rosenfield (2008) also called a 2006 conference on team science sponsored by the US-based National Cancer Institute a turning point in understanding the need for trans-disciplinary research in health across the globe.

Since the early 2000s, the emerging field of science of team science has been advancing a new form of crossdisciplinary and cross-sector work. Its claim to transcendence lies in generating new methodological and conceptual frameworks for analyzing social, economic, political, environmental, and institutional factors in health and well-being. A range of stakeholders are involved, including scientists, trainees, funders, policymakers, and clinical and community partners. The emphasis is on enhancing scientific discoveries and reporting evidence-based findings as well as improving educational outcomes, translational research aimed at clinical practices, and public policy (Hilton and Cooke, 2015). Not everyone agrees, however, on the nature of the field. Bozeman and Youtie (2017) acknowledged increased teamwork, complexities of collaboration, and strategies that foster success. However, they treated team science as a subset of collaboration studies. This book also adopts a broader definition proposed by the Team Scholarship Acceleration Lab at the University of California–Irvine. It acknowledges a larger ecosystem of teamwork both within and across sectors (https://tsal.uci.edu/).

Responding to Bozeman and Youtie, Kara Hall (2017) charged they mischaracterized and dismissed research and resources in the International Network for the Science of Team Science. Stephen Fiore (2018) also cited a wide range of methods and refuted their claims that Consultative Collaboration Management is the gold standard for managing collaboration and that team science lacks a strong evidentiary base. Debate on the status of team science is yet another episode in the boundary work of definition. The question of what constitutes interdisciplinarity is an even larger example. Olga Pombo contended interdisciplinarity is essentially a cognitive or philosophical phenomenon (pers. comm., 11 March 2020). Others, though, contend it is about institutional structure, and Lury et al. (2018) argued it is fundamentally about methodology. Growing attention to processes of boundary crossing has heightened interest in methods, though not without critics. Katri Huutoniemi (2014) faulted methodological foundationalism for imposing uniformity and predictability to ensure rigor and professionalism, while Robert Frodeman called emphasis on methods a kind of uniform proceduralism that confers a false "patina of objectivity" (2014, p. 49). Frodeman acknowledged the value of transferable skill sets and rules of thumb, but questioned whether there is any generalizable methodology.

Two books, though, highlight methods. Bergmann, Knobloch, Krohn, Pohl, and Schramm (2010, 2012) identified a repertoire of methods, instruments, tools, and strategies, including hypothesis and model building, integrative assessment procedures, boundary objects and concepts, heuristics, research questions, artifacts and products, mutual learning, and stakeholder participation. New methods also emerge in the course of conducting crossdisciplinary and cross-sector work. McDonald, Bammer, and Deane (2009) focused more specifically on mechanisms and strategies for communication, including well-known examples such as Delphi technique, systems methodology, scenario planning, and forums for stakeholder participation such as citizen juries, consensus conferences, and consensus development panels. Methods are not rote formulas, however. Bergmann et al. (2010, 2012) cautioned even if teams adopt formal languages as the basis for communicating across boundaries, they must be adapted to contexts and configurations of disciplines and stakeholders. Furthermore, methods are not strictly singular.

Ana Teixeira de Melo (2018) likened multi-methodological practice to abduction. It is an intuitive way of reasoning that is amplicative and generative, inquisitive and creative. Interdisciplinary relationships, she added, may become abduction by highlighting or pinpointing gray or obscured areas, illuminating commonalities and differences, changing landscapes by creating different contexts, and incorporating aspects of other disciplines or fields to expand possibilities or to increase congruence. Darbellay, Moody, Sedooka, and Steffen (2014) linked abduction to the serendipity of surprise, while Huutoniemi (2014) elevated heuristics over fixed methods. Rules of thumb, guidelines from practice, and weighing of possibilities develop through trial and error. Existing approaches, she acknowledged, can generate new ideas. However, they must be adapted to particular problems within an ecological rationality of pragmatism. Huutoniemi also contended new approaches lead away from the vertical logic of Western reasoning, aligning methodology with critique.

Upendra Baxi (2000) was equally mindful of the momentum their combination creates. The concept of transdisciplinarity, he recounted, arrived at a moment of wider crisis in the discourse of human rights accountability, expanding demand for rethinking the nature and the power of traditional approaches. Privileged and dominant forms of knowledge established genres, protocols, canons, and formations that deprivileged other ways of knowing. In response, new modes of knowledge, discourse, and institutional frameworks were demanded across all sectors of academic, private, and public life. Baxi highlighted in particular gaps between Western and non-Western traditions, esoteric and organic knowledges, colonial and Indigenous traditions, official and people's knowledges. One of the transgressive purposes of transdisciplinarity thus became renouncing the logic of instrumentality by reinforcing a democratic imperative that advocated participation of all stakeholders and their expertise, not just conventional protocols and truth claims of disciplinary and solely academic conventions. Thus, transgression aligned with the multi-layered discourse of critique. It interrogates instrumental problem solving, the principle of unity, disciplinarity, and even interdisciplinarity.

Critique

Although problem solving is often linked with accountability, Barry and Born (2013) cautioned it is neither a unified nor a monolithic discourse. Democratic imperatives constitute another form of accountability beyond instrumental goals. During the 1990s, TD began appearing more often as a label for knowledge formations imbued with a critical imperative. Kaplan and Levine (1997) associated a New Interdisciplinarity in humanities with theoretical paradigms that interrogate not only the canon and curriculum but also the larger organization of knowledge and hierarchies that govern intellectual and political lives. Ronald Schleifer (2002) linked it with transdisciplinary or cultural study of social and intellectual formations that breach canons of wholeness. Interdisciplinarity also became a target in challenges to instrumental forms. And, Jean-François Lyotard (1979, 1986) called interdisciplinarity specific to the age of delegitimation and hurried empiricism, charging it serves users of a conceptual and material machine through techniques of brainstorming and teamwork. Comparably, Bill Readings (1996) aligned interdisciplinarity with transformation of the Western university into a transnational bureaucratic corporation.

Looking back on earlier critiques, Graham Huggan (2002) suggested the title of a 1998 book, *The Anxiety of Interdisciplinarity*, summed up mounting concerns about threats to traditional disciplinary boundaries, institutional identities, and intellectual autonomy. Jill Vickers (1997) called new developments a by-product of the four posts of positivism, structuralism, modernism, and colonialism. All four unsettled previously stable ideas about knowledge and power. Inevitably, problem-focused crossdisciplinary approaches became linked with Funtowicz and Ravetz's (1993) concept of postnormal science as well. It is transgressive in breaking free of reductionist and mechanistic assumptions about how things are related and operate as well as the expectation science delivers a single-best solution or final answer. The heightened discourse of problem solving also meant it became linked with the concept of wicked problems that are not isolated to particular sectors of disciplines. They are interdependent, driven by complex cause-effect relationships. They exhibit a high divergence of values and factual knowledge. And they are emergent phenomena with non-linear dynamics, uncertainties, high political stakes in decision-making, and divergent values.

The discourse of critique calls to mind Donna Haraway's (2016) concept of staying with the trouble. It is linked with critical theory in humanities and some practices of social sciences, as well as feminist theory. Staying with the trouble does not settle for accommodation and resolution in theory and in practice as well as institutional forms. Tahani Nadim (2018) associated it with crafting, not recovering, and inventing, not finding. Moreover, it is disruptive. Mike Michael (2018) associated disruptive practices with surprise and disorientation, dissent, and meta-problemizing. Troubling is also compatible with the concept of diffraction in science. In physics it means a change in the direction of waves when they encounter an obstruction or overlap. Leila Downey (2018) emphasized diffraction embraces contamination and entanglement, rather than treating them as obstacles. Andrew Pickering (2013) further suggested challenges to conventional practices are better conceived as antidisciplinary because they contest underlying assumptions. He cited the field of cybernetics. It fit poorly within existing institutional structures, finding a social basis in cracks and interstices. Furthermore, its center of gravity was not the academy.

Critique also emerged in other discourses that intersect with inter- and trans-disciplinarity, including aforementioned examples of postnormal science, wicked problems, and Mode 2 theory. They go beyond methodological iteration and instrument integration to promote rethinking, reimagining, and reformulating. Thus they break free of reductionist and mechanistic assumptions about how systems operate and the expectation science delivers final answers with certainty. Mode 2, in particular, is associated with Gibbons et al.'s 1994 proposal that a new mode of knowledge production is fostering synthetic reconfiguration and recontextualization. In contrast to Mode 1—characterized by hierarchical, homogeneous, and discipline-based work—defining traits of Mode 2 are complexity, non-linearity, heterogeneity, and transdisciplinarity. New configurations of research are being generated continuously, and new social distribution of knowledge is occurring as a wider range of organizations and stakeholders contribute skills and expertise. As boundaries of control blur, underlying notions of competence are redefined and new criteria are needed.

Gibbons and co-authors initially highlighted instrumental contexts such as aircraft design, pharmaceuticals, electronics, and product development. In 2001, however, Nowotny, Gibbons, and Scott extended Mode 2 theory. They argued contextualization of problems requires participation in the *agora* of public debate. When lay perspectives and alternative knowledges are recognized, a shift occurs from solely reliable scientific knowledge to inclusion of socially robust knowledge that dismantles the expert/lay dichotomy while fostering new partnerships between the academy and society. Scientific knowledge continues to have a role, but not a privileged one. This imperative was not new and anticipated current ascendancy of problem-oriented transdisciplinarity. O'Brien, Marzano, and Whit (2013) traced the concept of stakeholder participation to theories of deliberative democracy. During the 1970s calls were heard for co-management of renewable resources and environments, and Kessel and Rosenfield (2008) reported the discourse of problem solving was taking hold during the 1990s in both health and social sciences in the development world. By the late 1980s and 1990s new ideas for planned participation in environmental regulation had emerged as well. Growing momentum for both collaborative problem solving and transdisciplinarity set the stage for wide adoption of Mode 2 theory, but it is not without blind spots.

Weingart (2000) cited failing to recognize the opportunism of academic capitalism, overstating theoretical claims, and falling short of transformation. Moreover, he added, both Mode 2 and postnormal science often look at phenomena on the surface, describing institutional changes rather than advancing a new epistemology. Current discourse of inter- and trans-disciplinarity, Weingart concluded, exhibits a common topos among claims for new modes of knowledge production. They oscillate between empirical and normative statements situated in contexts of application and stakeholder participation. Yet, here too efforts fall short of transformation. Merritt Polk (2015a) contended the goal of making science more accountable cannot be met within standard parameters. It must become subversive in three ways: practically, by disrupting established processes; epistemologically, by challenging conventional criteria of reliability and rigor; and contextually, by forging a neutral space between science and practice.

Bridging Discourses

Polk's call clearly crosses boundaries of discourse. However, Graff contended a standard version of interdisciplinarity prevails, singling out the US National

Institutes of Health (NIH) as an example of a dominant narrative aligned with problem solving. However, NIH is a large federation of institutes and does not follow a single definition. Multiple constructions co-exist, not only in the context of health research but in other areas as well. Mark Kann's (1979) classification of three political positions provides a reference point for understanding why discourse and related positions are simultaneously distinct yet also intersect. The first—Conservative Elites—prioritizes instrumental solving of social and economic problems without disrupting the status quo. The current rhetoric of convergence, though, walks the line between problem solving and epistemology while beckoning transformative change in an accompanying critique of limits of existing structures and practices. Convergence has become a concept *du jour.* Like others, though, it has multiple connotations. It appears in conjunction with creating greater coherence across a campus around cross-secting themes often framed by the discourse of problem solving in addressing grand challenges of the day. It is also a process central to Chapter 4: building collective intelligence in groups.

Weinberger, Stegmann, and Fischer (2007) conceptualized knowledge convergence in the latter connotation as three stages. The first—Sharing—occurs when learners possess the same concepts. The second—Equivalence—connotes a similar degree of knowledge about a subject prior to collaboration. The third—Convergence—is a higher level developed through collaborative learning and transactive relations. In yet another example, convergence labels a prominent trend at intersections of life sciences, physical sciences, medicine, and engineering. Sharp and Langer (2011) deemed it the latest of three revolutions. The first revolution was molecular and cellular biology, and the second genomics. Both advanced interdisciplinary research on biological processes. However, the third revolution beckons a trans-disciplinary framework for problem solving through cross-sector collaborations. It is the focus of a state-of-the-art report on convergence, which called the third revolution an "expanded form of interdisciplinarity" (NASEM, 2015, p. 20). The report tallied prominent goals: understanding complex biological systems, improving patient outcomes, revolutionizing manufacturing, enhancing energy storage systems, and providing secure food supplies. Problem solving is clearly an imperative, but the movement has philosophical implications as well.

The underlying dynamic is a process of divergence and convergence linked with the concept of combinatorial innovation. It has both theoretical and methodological implications. The process occurs, Roco, Bainbridge, Toon, and Whitesides (2013) explained, when a new technology or set of them yields components that may be combined and recombined, spinning off applications that continue to be recombined and integrated. In the mode of transdisciplinarity, this connotation of convergence aims to create a comprehensive synthetic framework by merging knowledge from multiple fields and ideas into macro-domains of research activity, with promises of new inventions, innovations, treatment protocols, and forms of education and training. Yet, Roco et al. cautioned a new broad-based knowledge network has not materialized. Research and development have been more reactive and co-incidental than transformative.

The second position Kann identified—Radical Dissidents—parallels the discourse of critique. Radical Dissidents prioritize needs and problems of oppressed and marginalized groups in a socio-political agenda. Critique alone is not enough, however. Popa, Guillermin, and Dedeurwaerdere (2015) argued it does not guide change through pragmatic actions that emerge from agreement on a normative pathway. When defining issue-driven interdisciplinarity, John Robinson (2010) highlighted claims of practice over theory and epistemology. And, Sue L. T. McGregor (2020) linked transdisciplinary transpraxis with action, not theory and philosophy. Yet, transdisciplinary transpraxis shares with postmodernism concern for binaries and dualisms, hierarchies and power. Thus, it critiques belief and value systems, ways of thinking, and structures of understanding. McGregor also aligned it with emergence, relationality, and contingency. Others are explicitly bridging discourses of philosophy and problem solving. Scholz et al. (2011), for example, argued socially robust knowledge involves a form of epistemics that bridges scientific and experiential knowledge, while Polk contended, "Practical usability is both epistemic and non-epistemic" (2015b, p. 113).

Here too, implications follow for epistemology. Arnim Wiek called interactive knowledge generation in transdisciplinary research a "third epistemic way" (2007, p. 52), and Hans Dielman (2015) treated transdisciplinary hermeneutics as more than a simple sum of experiential, formal, and direct knowing. Dielman called transdisciplinary hermeneutics a multiplier that facilitates emergence in spaces or platforms of imagination and experimentation, thus bridging philosophy, problem solving, and critique. Further implications follow for the future of interdisciplinarity. Frodeman (2017) aligned the concept with the urgency of a transdisciplinary commitment to problem solving, while Machiel Keestra (2019) called for greater commitment to a socially robust connotation. It still requires metacognition and reflection, however, in order to grapple with multiple viewpoints. Moreover, Keestra added, involvement of stakeholders outside the academy pushes beyond defining interdisciplinarity solely in relation to disciplinarity while prioritizing action research guided by both a democratic imperative and a philosophical grounding in complexity and emergence.

Meyer and Peukert (2020) traced warrants for the underlying problematic to critique of the subject/object dichotomy in 20th-century French and Belgian philosophy. Gaston Bachelard treated it as an epistemological break from universal assumptions. Gilbert Simondon's onto-epistemological concept of individuation emphasized a situated, open process of being and becoming. Isabelle Stengers' concept of an ecology of practices underscored particularity and context. And, Patrice Maniglier's conception of designing as a creative practice paved the way for a transformative epistemology. Meyer and Peukert combined these ideas in the concept of a new thinking practice of problematic design for trans-disciplinary sustainability research. Bridging epistemology and problem solving also opens up a more expansive view of what McGregor (2015) dubbed the Zurich (Swiss) approach to transdisciplinarity, which she contrasted to a Nicolescuian approach.

Both acknowledge complexity and emergence while conceiving of unity within plurality, although the latter is associated with Nicolescu and the Centre International de Recherches et Études Transdisciplinaire. Founded in 1987, the Center is advancing Nicolescu's philosophical and methodological approach in a vision of transdisciplinarity that replaces reduction with a new principle of relativity. It is not only transcultural and transnational but also encompasses ethics, spirituality, and creativity. In contrast, McGregor purported, the Zurich approach synthesizes knowledge of disciplines and social actors in order to foster socially robust, reflexive, and accountable research without concern for reality, axioms, or logics. Thus, she contended, it privileges doing science better. However, the classification Zurich Approach is narrow, ignoring epistemic and ontological dimensions of co-production of knowledge (Hirsch Hadorn et al., 2008). Furthermore, it emerged during the late 20th century in other German-speaking countries besides Switzerland, in Austria and Germany, and some related interests in the Netherlands and Sweden. By now, it is also an influential concept in literature in Australia, Latin America, and Africa.

The third position Kann identified—Liberal Academics—demands accommodation but retains a base in existing structure, in the middle ground of harmonious interactions. In classifying types of institutional change, Toombs and Tierney (1991) acknowledged tactics of a first model—Modification—can add knowledge, techniques, and practices. However, connections are often limited to special programs. The prevailing structure, purpose, and characteristics of an institution remain intact. A second model—Integration—can foster connections but is often partial, and interdisciplinary programs are only tolerated. Hence, as we shall see in the closing chapter, they are more vulnerable. In contrast, a third model—Transformation—recognizes the need for new priorities, including interests that form the basis of new fields in Chapter 3, such as gender, multiculturalism, and globalism. This model is more likely to foster Wátzlawick, Weakland, and Fisch's (1974) notion of second-order instead of first-order change they likened to simply moving the furniture around.

A 2011 volume entitled *Transdisciplinary Knowledge Production in Architecture and Urbanism* illustrates the potential for second-order change in professional practice that also bridges philosophical warrants, problem solving, and critique. Editors Doucet and Janssens (2011) challenged strict separation of discourses. New hybrid modes of inquiry, practice, and learning have the capacity to overcome past schisms of theory, history, and practice as well as critical theory and projective design. TD also places ethics, aesthetics, and creativity inside disciplinary and professional work rather than marginalizing them to external interests. Social and political, normative and ethical questions are thereby incorporated; new objects come into view; practices are placed in new configurations; and theory and learning are contextualized and resituated. Awareness of hybridization also heightens when incorporating once excluded forms of knowledge, including understandings of lay people. In addition, the current heterogeneity, relationality, and hybridity of knowledge is heightened.

Writing in the same volume, Tony Fry (2011) called transdisciplinarity a conceptual leap beyond pre-made methods. It is not simply a matter of synthesizing existing methods. "Problems are never received," he explained, "but always interrogated and redefined"; in short, "architecture never just begets architecture" (p. 21). It becomes a form of redirective practice that creates new ways of dealing with complexity. Bryan Turner's (1990) analysis of the medical curriculum also returns to mind when thinking about the boundary work of redirecting practice. Pragmatic questions of reliability, efficiency, and commercialism take center stage when interdisciplinarity is conceived as a short-term solution to economic and technological problems. In contrast, it emerged in social medicine and sociology of health as a philosophical goal focused on the complex causality of illness and disease. Researchers developed a holistic biosocial or biopsychosocial model critical of the traditional hierarchical model of medicine.

The earliest examples—social medicine and sociology of health and illness—bridged discourses of epistemology and critique of segmented disciplinary assumptions about medical problems, interventions, and training. Although they arose from different social and historical circumstances, they shared the concept of the whole person and the multi-causality of illness, including biological, social, economic, cultural, and political factors. A critical epistemology, Turner emphasized, is further linked with a shift in the nature of disease from acute illness to problems with a multifactorial etiology and social component, such as cancer, heart disease, cerosis, and arteriosclerosis. The first two examples were limited by the positivist model of medicine, while the third and fourth examples faced other limits. The third is problem-oriented interdisciplinary research centers. Corporatization of healthcare has resulted in commercializing knowledge and practices within industrialized nations, propelling a shift from the reputational model based on specialization and autonomy of medical professionals. Pragmatics of management and funding take priority over both philosophy and critique. The fourth—postmodernism in social theory—would seem to align with critique, including decline of meta-narratives and erosion of traditional hierarchies and divisions of labor. Yet, Turner cautioned, commercialization of medicine and postmodern critique of the medical curriculum may converge, resulting in further fragmentation as research units proliferate.

Whose Knowledge Counts?

Comparing the three discourses while also recognizing their intersections raises a final question: Whose knowledge counts? Looking ahead to Chapter 3, interdisciplinary fields that emerged from socio-political movements outside the academy created platforms for marginalized groups. Demographic shifts, Louis Menand (2001, 2010) noted, coincided with cultural critique in the late 1980s and early 1990s. Their cumulative effect reinforced calls for greater inclusion of a wider array of perspectives and forms of knowledge. On a broader international scale, Hirsch Hadorn et al. (2006) cited a Southern version of transdisciplinarity shaped by theories of social action and transformation, including Brazilian author Paulo Freire's concept of the pedagogy of the oppressed and Leonard Boff's concept of liberation of the oppressed. Along with participatory rural appraisal they challenged models of economic and technological development centered on discipline-based scientific knowledge. As a result, social and political indicators gained priority over income, yield, technology, and other measures of modernization. They too reinforced calls for greater recognition of diversity and ways of knowing often marginalized by dominance of academic knowledge.

Knapp, Reid, Fernández-Gimenéz, Klein, and Galvin (2019) included Indigenous and local knowledge in a roster of transdisciplinary collaborative approaches encompassing traditional knowledge (passed through generations), empirical expertise (based on careful observation and practice), revealed wisdom (transmitted in vision, ritual, and ceremony), and contemporary perspectives (drawn from experience, education, and problem solving). These and related concepts and movements come together in efforts to decolonize transdisciplinarity. Yet, reflecting on projects in Australia, Michael Davis (2004) asked whether efforts to incorporate both Indigenous knowledge and languages of official law and policy are bridging a gap or crossing a bridge. Western intellectual traditions are still often privileged as rational, rigorous, and technical while local, traditional, and Indigenous traditions and discourses are deemed intuitive and informal. Davis is not alone in his critique. Davora Yanow (2004) noted expert knowledge prioritizes theory, abstraction, scientific construction, academic and scholarly foundations, and professional concerns. In contrast, local knowledge is based in practice, interaction, lived experience, practical reason, the tacit, and the everyday. Moreover, Yanow added, "[l]ocal knowledge is contextual knowledge" (p. S11).

The gaps are recognized in multiple countries in efforts to assert other worldviews, thereby enlarging possibilities for addressing challenges being faced at local, national, regional, and international levels. In making the case for decolonizing TD in Africa, Bagele Chilisa (2017) explained that methodologies rooted in local philosophies, worldviews, and history contribute alternative ways of conducting research. Ethno-philosophy, for example, focuses on how mythical concepts, ritual practices, language, proverbs, metaphors, folklores, stories, songs, artifacts, and art can inform epistemology and methodology, maximizing connections, participation, and relationality. Du Plessis, Sehume, and Martin (2013) linked transdisciplinarity with attempts to combine concepts from Western philosophy of science and African cosmology in the Mapungubwe Institute. Eden, Donaldson, and Walker (2006) also reported that NGOs are adapting elements of classical science to their own purpose in a Mode 2 model incorporating principles of democratization, contextualization, and utility.

Mode 2 theory has gained recognition in higher education policy in South Africa as well, though Mercy Brown-Luthango (2013) reported influence has been small and uneven. Launched in 2011, the Mapungubwe Institute for Strategic Reflection situates discourse of transcendence within the context of problem solving, including efforts to address challenges of climate change, poverty, and unemployment. At the same time, it transgresses Western precedents even while mindful of the Western lineage of philosophers and social theorists as well as complexity theory. Thus, transdisciplinary approaches to sustainability can bridge not only divides of epistemology and instrumentality but also producers and users of knowledge in the North/West and Global South. Sustainability research critiques the top-down model of knowledge transfer and the colonial paradigm of capacity building signified by economic metaphors of banking concepts and deficit models.

Comparably, when introducing a collection of articles on interdisciplinarity in Latin America, Bianca Vienni Baptista (2016) proposed regionalism is a defining concept in that area of the world, in the dual sense of systematizing lessons across a network of countries and grounding both pragmatically and philosophically in real-world problems, including challenges of development. In defining decolonizing practice in Latin America, Danilo Streck (2007) has linked inter- and trans-disciplinarity with developing an alternative to traditional forms of knowledge production, calling it a border praxis without a fixed location. Building on Streck's work, Machiel Keestra (2019) highlighted Orlando Fals Borda's *Investigación Acción Participativa* (participatory action research), Paulo Freire's *Pesquisa Participante* (participatory research), and *Sistematización de Experiencias* (systematization of experiences). Freire recognized dynamics of cross-boundary participation, while *Sistematización de Experiencias* seeks alternative criteria of relevance and validity grounded in neighborhoods, communities, and social movements. Actionability, Keestra added, becomes a criterion for interdisciplinary knowledge. Action research implies knowledge is valid from different perspectives and relevant in a social context required by transdisciplinarity.

Bridging theory and situated practices questions the dichotomy of general abstraction and specificity of cases. Wolfgang Krohn (2017) contended the relationship of the nomothetic and the idiosyncratic is nothing less than the epistemological challenge of interdisciplinarity. The nomothetic tends to seek abstract laws and theoretical frameworks while the idiosyncratic often accentuates pragmatism framed by particular scientific and societal problems. Crossing their boundaries demands a new mode of knowledge that not only bridges discourses of philosophy and problem solving. Krohn emphasized it also combines humanistic understanding of particularities and contingencies of cases with a scientific search for universal, abstract, and law-like principles. Case studies, he added, are not only expected to solve single problems. They also contribute to stocks of knowledge, even though their epistemic structure differs from general theories or paradigms. In an earlier reflection on discourses of transdisciplinarity, I identified three positions that overlap at times (Klein, 2014). They merit revisiting in light of this chapter's focus on discourses of philosophy, problem solving, and critique with updated examples, new reflections on the boundary work of positioning them, and added emphasis on bridging discourses.

The first—transcendence—was initially epistemological without acknowledging the role of ontology as well. It did not factor in new thinking about a transdisciplinary epistemic of pragmatism, either. In describing the discourse of problem solving, it did not take into account recent reports from authoritative bodies, including mounting calls for transdisciplinary rather than older interdisciplinary approaches and new scholarship in related networks. The third—transgression—also needs updating to include bridging discourses of both philosophy and problem solving. All three discourses, it should be added, bespeak transformative change. However, change is a complex process. In a book about change in a wide array of contexts, Howard Gardner (2004) concluded it results most often from a slow shift of viewpoints rather than a single argument. Changes that seem to erupt suddenly also mask subtle processes that gel over a longer period. And, it is more difficult to replace a simple way of thinking with a more complex one.

To no surprise, then, resistance to change is widespread. The British Academy's (2016) report on interdisciplinarity, for example, urged individuals to have a departmental base. Calling it a normal home, however, reinscribes institutional hierarchy. The report of a workshop on Fostering a Culture of Convergence noted that researchers and agencies addressing biomedical, agricultural, energy, and ecosystems challenges were included in meetings. However, the Department of Transportation was not, even though transportation systems must address related concerns. Computational sciences and social sciences were also relatively isolated from conversations (NASEM, 2019). As a result of resistance, tradeoffs are inevitable. Kinchy and Kleinman's (2003) description of the Ecological Society of America's boundary work targeted four areas: historical discourses of value-freedom and utility, national politics, structure and actions of other societies, and orthodoxies of scientific behavior. Yet, the Society's effort to balance discourses of purity and pragmatism varies as political contexts change. Moreover, fields respond to both internal and external stakeholders. Echoing Burton Clark's (1995) findings from an international study of research universities, fields also exhibit unique organizational characteristics that enable them to meet multiple, ill-defined, overlapping, and sometimes conflicting goals that are responsive to both internal academic and external work on complex problems and questions in multiple directions from traditional academic settings.

Zietsma and Lawrence's (2010) study of the boundary demarcating the forestry field in British Columbia offers a final example, with two generic insights. It did not take place in an academic setting. However, parallels to both crossdisciplinary and cross-sector work appear, especially the social nature of boundary work. It occurs between clusters of individuals and groups. In contrast to boundary work, practice work creates, maintains, and disrupts practices considered legitimate in a particular field. The second insight is a scheme of four cycles of stability or change that also appear in the next chapter on new interdisciplinary fields. Stability involves maintaining practices and their concomitant boundaries. Conflict leads to breaching or bolstering them, thereby disrupting or defending practices. Innovation, in turn, can generate new practices, and restabilization can occur when making connections across boundaries and diffusing practices. Thus, bounding is subject to change in an ecology of spatializing practices at different levels of formality and complexity. It also involves different social actors, is influenced by both endogenous and exogenous factors, and results in different degrees of success and failure elaborated further in the closing chapter of this book on failing and succeeding.

In their book on *Interdisciplinary Discourse,* Choi and Richards further cautioned that "hard research" into the nature of interdisciplinary research is still too limited to substantiate claims about how it works (2017, p. 253). They conceded literature is rich in accounts from experience and pragmatic recommendations emanating from them, as well as theorizing and typologizing. However, they argued the topic is still neglected. The next three chapters, though, dig deeply into its workings. In doing so, they do not abandon discourse analysis. Patterns of argument and use of terminology remain important to understanding how theories and practices are positioned within the larger landscape of knowledge production both within and beyond the academy and at all levels, from small-scale projects to networks to fully formed communities of practice. The chapters also continue to prioritize the written word while taking into account evidence from the spoken word in interview-based studies, analysis of transcripts, and feedback in workshops.

Part I of this book ends in Chapter 3 by examining both the what and the how of building and sustaining new interdisciplinary fields. Here too it walks the lines of generality and specificity by identifying both similarities and differences in how fields operate while emphasizing their heterogeneity. In keeping with greater emphasis on pragmatics, Part II examines their forms of argument. Chapter 4 examines challenges and dynamics of communicating that lie at the heart of collaborating, developing in the process of dialogue extending from initial articulation of individual positions to formulation of shared language cultures. Chapter 5 focuses on the how of capacity by depicting the centrality of learning in both crossdisciplinary and cross-sector work along with related competencies.

3
Interdisciplinary Fields

> Another distinct type of interdisciplinary program focuses on niche knowledge areas, such as gender studies, environmental studies, race/ethnic studies, and human ecology. Each field represents unique influences in terms of its development.
>
> **Karri Holley (2017)**

Kinchy and Kleinman (2003) distinguished strategic boundary work that advances targeted stakes from routine boundary work. The former is often small-scale and short-lived, whereas the latter is a longer-term endeavor on a larger scale. This chapter investigates the second type of boundary work in communities of practice known as interdisciplinary fields and interdisciplines. The distinction between the two labels, however, is not absolute. The most in-depth study of interdisciplines for instance, Harvey Graff's (2015) *Undisciplining Knowledge*, includes both terms and sometimes uses them interchangeably. Recalling the preview of this chapter in the Introduction, Richard McKeon explained new fields emerge because subject matters are not ready made to respond to all questions, problems, and issues that arise. Interdisciplinarity is thus an "architectonic art" of creating new formations (1979, p. 18). As we shall also see, echoing Lynton Caldwell's (1983) contention, the metaphor of fit prejudges the epistemological question at stake in formation of fields that arose because of misfit of needs, experiences, and information with disciplinary structures.

The political economy of the academy, Stanley Katz (1996) acknowledged, forces new fields to play a traditional power game of seeking resources and securing institutional space. However, they mark the possibility of a more flexible design. Thus their boundary work is two-fold: walking a line between securing an institutional foothold while advancing new interests that do not fit into conventional categories and niches. This chapter begins by describing catalysts for new fields, then draws insights from patterns of interdisciplinary majors in higher education as well as taxonomies of both research and education. It next compares trajectories of particular fields and turns after that to

Beyond Interdisciplinarity. Julie Thompson Klein, Oxford University Press. © Oxford University Press 2021.
DOI: 10.1093/oso/9780197571149.003.0004

plural identities for both individuals and fields while illustrating the boundary work of naming them. The chapter concludes by addressing multiple themes through the lens of women's studies and intersectionalities of fields today, ending with the question of whether there is a distinctive interdisciplinary logic. In the course of discussion, it returns to earlier accounts to update and situate them in current complexities of boundary work. Comparison over time sharpens understanding of why fields do not fit a universal formula and evolve to respond to new needs and interests.

Nature of Interdisciplinary Fields

When Raymond Miller (1982) classified crossdisciplinary efforts, he identified four categories that are major catalysts for new fields. The first—Topics—formed the basis for many of them, including area/crime/environmental/and urban studies as well as aging in gerontology. The second—Life Experience—became a prominent catalyst in the late 1960s and 1970s with emergence of fields that in many cases emanated from socio-political movements outside the academy, including Black/ethnic/and women's studies. Douglas Bennett called them a "sacred edge" in the reopened battle over inclusion and exclusion (1997, p. 144). By encompassing conventions of cultural and postcolonial studies they challenged notions of class, gender, race, ethnicity, and identity, and they bolstered the democratic imperative in the discourse of critique. The third—Professional Preparation—is linked with occupational fields, such as law, engineering, social work, and nursing. Neil Smelser (2004) also cited organization and management studies as well as planning and public policy. The fourth—Hybrids—refers to interstitial specializations or interdisciplines such as historical sociology, medical anthropology, social psychology, and molecular biology.

Boundary consciousness appears in all four categories. Proponents have differing commitments to historical versus contemporary approaches, internalist versus externalist priorities, quantitative versus qualitative practices, and discourses of philosophy, problem solving, and critique. Two studies yield a more detailed empirical picture of the institutional foothold of fields. The Social Science Research Council (SSRC) combined in-depth interviews with faculty and administrators of eight campuses with a survey of 222 liberal arts institutions in the US, aimed at mapping patterns of interdisciplinary education. Of the top 10 majors, more than half were in international, global, or area studies; two followed a biology-plus model; and the most popular fields were associated with advocacy and activism. The most frequently cited examples were environmental studies and science, women's and gender studies, neuroscience and psychobiology, and American studies. Reflections on future trends were equally insightful. Two-thirds of respondents to the survey expected offerings to increase. The most commonly cited reason was research, based on a belief that questions students and faculty are investigating today often require expertise of scholars from more than one field. Student interest in new areas was an added factor (Rhoten, Boix-Mansilla, Chun, and Klein, 2006).

In a rare longitudinal study of 293 institutions extending from 1975 to 2000, Brint, Turk-Bicakci, Proctor, and Murphy (2009) presented a fuller though parallel picture of undergraduate degree programs. They found the growth rate of interdisciplinary majors in the US was strong. Units tripled in international relations, global studies, and fields of race, ethnic, and women's studies as well as brain and biomedical science. Environmental studies nearly tripled, while cultural studies beyond the West more than doubled. Western studies also nearly doubled, though civic/governmental studies and American studies grew more slowly. Despite evidence of growth, however, the rate is uneven. In an earlier reflection on forces propelling universities to move in new directions, Steven Brint (2005) called attention to variances between sciences and humanities that affect the character of fields. The discourse of problem solving is strong in sciences, including economic ties. In contrast, democratization is a strong driver in humanities.

Two additional reports yielded further documentation. When a subcommittee evaluated the US-based National Research Council's taxonomy of research-doctorate programs, its members recommended increasing the number of recognized fields from 41 to 57. They also recommended the category of global area studies be expanded and the discipline of biology be renamed life sciences, while including agricultural sciences and merging mathematics and physical sciences with engineering (Ostriker and Kuh, 2003). The final report highlighted life sciences. Appendix C, however, acknowledged emerging fields of bioinformatics and biotechnology, computational and nuclear engineering, and nanoscience and nanotechnology. In addition, systems biology and information science appeared as well as race/ethnicity/postcolonial/feminist/gender/and sexuality studies. The Appendix also recognized film studies, criminology and criminal justice, science and technology studies, urban studies and planning, and rhetoric and composition (Ostriker, Holland, Kuh, and Voytuk, 2009). An even more extensive picture appears from the US Department of Education's National Center for Educational Statistics, which classifies both undergraduate and graduate programs.

In 2005 the National Center's classification Multi/interdisciplinary studies (MD/ID) encompassed a general category, syntheses of biological and physical sciences, mathematics and computer science, and "studies" programs. Beyond MD/ID, though, two other classifications revealed patterns in added accounts: a composite code including area/ethnic/cultural/and gender studies and a combination of disciplinary and interdisciplinary studies in biological/biomedical sciences. Comparison of 2000 and 2010 versions of the companion Classification of Instructional Programs documented new developments in healthcare, including a "multidisciplinary" Child Abuse Pediatrics Residency Program, and within "interdisciplinary" programs a Clinical Nurse Leader on a team, Human Biology, and Pediatric Rehabilitation Medicine Residency Program. Molecular Medicine and Neurobiology had new designations as well, and older categories were renamed to reflect expansion. For instance, Information Technology became Infomatics. The 2020 update, in turn, added 300 new programs, including Human-Centered Technology Design in Infomatics, STEM in Early Childhood Education and Teaching, and new positions in nanotechnology (https://nces.ed.gov/ipeds/cipcode/crosswalk.aspx?y=56).

Taken together, these studies and taxonomies document increases in both biomedical and cultural fields. However, when prompted about future directions, respondents to Brint, Turk-Bicakci, Proctor, and Murphy's survey predicted brain and biomedical science fields such as cognitive science and neuroscience were more likely to be institutionalized in departments. This belief reflects the heightened priority of problem solving and economic imperatives over discourses of philosophy and critique. Insightful as they are, however, rosters of examples are only blunt measures. They do not furnish an in-depth picture of the boundary work of constructing and sustaining new fields. Charles Briggs' (2008) account of efforts to build folkloristics illustrates the role that asserting professional authority plays. Richard Dorson drew a divide between legitimate folklore from fakelore and frivolous work. This kind of discipline-building, Briggs added, fostered commitment to a new professionalized domain and an *esprit de corps* in this, and as we shall see below, other fields.

Trajectories

Economic and social capital are strong factors in growth and sustainability of interdisciplinary fields. Area studies and molecular biology, for example, were buoyed by higher levels of support than social psychology and cultural studies. Area studies received significant funding from the Ford Foundation, and molecular biology continues to enjoy widespread support for life science research and its biomedical benefits. By extension, it should come as no surprise that terrorism studies and pandemic-related research now have greater priority than performance studies and digital humanities. Institutional foothold is a key factor. In physics the term *critical mass* refers to the minimum amount of nuclear fuel needed for a chain reaction to start. As a metaphor for new fields it signifies minimal resources needed to escalate a shared interest. They include an adequate number of dedicated scholars and educators, the scale of research and educational programs, financial and other resources, an infrastructure for communication, a scholarly body of knowledge, and a self-defined identity (Klein, 1996). Harvey Graff's (2015) comparison of interdisciplines sheds light on why fields do not attain the same level of critical mass and criteria for judging success or failure.

Graff deemed social relations a failure because it did not meet a stated goal of unifying social sciences and was limited to a single institutional place, the Social Relations Program at Harvard University during the 1930s to 1960s. In contrast, he extolled operations research as an exemplary form of interdisciplinarity. Integration and synthesis were anchored in problems and questions, in contrast to the dominant influence of one individual in Social Relations, Talcott Parsons. In addition operations research was cross-sector, bridging pertinent disciplines as well as professions, industrial engineering, hybrid groups, and think tanks outside the academy. Graff also judged materials science during the 1950s–1990s a success. Institutional spaces provided a strong infrastructure, in this case the professional organization of the Materials Research Society and the Center for Materials Science and Engineering at MIT. Graff considered cultural studies a failed example as well. His dismissal of its amorphous character, however, is debatable. He presented a number of definitions, including an entry in Wikipedia. However, he bypassed a large body of scholarly literature that provides more in-depth understanding of the field's unique character, calling into question universal criteria for determining failure and success.

David Bathrick (1992) called cultural studies a gravitational field or magnet that gathers coalescing practices into a network of overlapping projects that crossfertilize rather than settle on one formation. As a result, studies of culture are located within disciplinary subfields though also operate at their margins and in alternative practices. Thus the field is multidisciplinary in scope, interdisciplinary in situated practices, and transdisciplinary in promoting a new framework for studies of culture. Graff was dismissive of these distinctions. However, they are important signposts of multiple practices even in the same field. Furthermore, cultural studies is counterdisciplinary in Bathrick's sense of a problematizing force (Klein, 1996) rather than a single construct. Identity is also linked with geographical context. Celia Lury (2018) cited the concept of an Inter-Asia method and an intellectual movement of Inter-Asia Cultural Studies.

In addition, Graff judged cognitive science less successful. He traced the origin of a science of mind to opposition to the behavioral science movement and conventional political and intellectual history. Graff concluded cognitive science defined interdisciplinarity by merely listing disciplines rather than examining interrelationships or core questions, deeming it a particularized form of multi- or cross-disciplinarity with a presence in fields such as artificial intelligence, neuroscience, and, to a degree, cognitive psychology. Not everyone agrees, however. Paul Thagard's (2017) updated account for *The Oxford Handbook of Interdisciplinarity* provides an extended example, portraying the field as a thriving area of research and teaching. The term *cognitive science*, he recalled, was not used until the 1970s. However, the field has roots in the 1940s, in attempts to apply emerging ideas about computing, engineering, and brain systems in order to develop new hypotheses about mental mechanisms. Over time cognitive science gained an institutional foothold with dedicated academic programs and the professional anchoring of a learned society, a dedicated journal, and body of scholarship. Yet, Thagard acknowledged, disciplines have been unequal participants.

Despite contributing insights into mental representations and processes in non(*sic*)-Western cultures, for example, most anthropologists have not been interested in cognitive science. Psycholinguistics flourished early on, but neuroscience gained stature in the 1980s and 1990s. Styles of connection vary as well. In one type individuals conduct research at intersections of two or more disciplines. Psychologists, for instance, engage in computational modeling, and some philosophers perform experiments in psychology or neuroscience. Another type entails collaborative research, while yet another draws on ideas from related fields. Plurality is also evident in publications. Many articles in the journal *Cognitive Science,* Thagard acknowledged, are not explicitly interdisciplinary, but they foster breadth. An experimental essay on human concepts that fits into cognitive psychology, for instance, is pertinent to philosophical, computational, neurological, linguistic, and cross-cultural questions about mental representations.

Some textbooks divide descriptions into philosophy, psychology, neuroscience, linguistics, and artificial intelligence. Nonetheless, Thagard highlighted two areas that validate the claim of interdisciplinarity. The first, theoretical development, is evident in new ideas that bridge representational structures and computational procedures. Psychology has drawn on ideas from philosophy and linguistics, while making contributions to artificial intelligence. Some expert systems have benefited from psychological ideas about mental representations. And, developments in cognitive psychology have influenced work in philosophy of mind and cognitive anthropology. The second area, experimentation, includes new ways of testing theoretical ideas and measurement tools, including computer technology. Linguistics has also furnished theoretical ideas about rules and representations. In addition, philosophical ideas have led to psychological experiments on prototypes, and neuroscience is fostering new efforts in cognitive psychology, amplified by tools for measuring mental activity. Factoring in both theoretical development and experimentation, Thagard highlighted five key factors for success of any interdisciplinary field explored further in the closing chapter of this book; they include ideas, methods, people, places, and organizations.

Change

As the foregoing examples demonstrate, historical perspective is crucial. In cognitive science, for instance, the subarea of artificial intelligence has moved from interest in human thinking to an engineering-oriented focus of a quantitative bent. One of the earliest fields, area studies, has also changed over time. It was initially a by-product of World War II, centered on US interests. It was multi-layered, though. Richard Lambert (1991) described it as not only variegated but fragmented, rather than a homogeneous tradition. Initial work occurred mostly at the juncture of four disciplines: history, literature and language, anthropology, and political science. At that hybrid space, he recalled, a historically informed political anthropology emerged using material in local languages. The dominant pattern was broadly defined themes, creating multidisciplinary perspective on a particular event. In an updated genealogy, Craig Calhoun (2017) recalled different emphases in the early years. The Cold War meant politics was at the center of Russian and Eastern European studies. South Asian studies recognized political concerns as well but focused more on civilization and culture. In contrast, economic development was prominent in Latin American studies, and formation of new nations in African studies. Other factors also led to change.

The first major paradigm—modernization—declined with diminishing support for Western approaches to economic development and modernization theory. Drops in funding in the mid-1970s and pressure to establish disciplinary credentials curbed expansion, and a long-standing epistemological fault was evident as well. In contrast to the search for general laws in disciplines, area studies were particularizing, focused on local conjunctures of history, culture, politics, and environment that fostered crossdisciplinary discussions of underdevelopment, class and power, states and nations. From the 1970s forward, many researchers began to conceptualize globalization as the basis for a universal theory. Today, however, Calhoun reported area studies are being reimagined with new connections between regions and geopolitical areas once considered underdeveloped. Canadian studies offers a parallel lesson about change.

Jill Vickers (1997) recalled Canadian studies was once considered a single kind of activity. Now, however, multiple fields are "in the mix." Some have "congealed" to the point they have a recognized canon or foundational theory, anchored by journals, learned societies, and, in some cases, departments. In contrast, "open or cross-roads interdisciplinary fields" do not necessarily settle on a canon, foundation, paradigm, or epistemology, including identity fields that emanated from pressures for societal change such as women's, Quebec, and First Nations' movements. The field has a firmer presence now. Yet, Vickers found, multidisciplinarity persists in disciplinary-based training and in books and conferences. Even so, a generational dynamic reported in other fields is also apparent. By the late 1990s more than half of the students in her graduate courses had interdisciplinary undergraduate degrees and, increasingly, junior instructors and teaching assistants had experience in inter- or transdisciplinary fields (pp. 15, 21). South of the border separating Canada and the US, American studies offers another parallel that illustrates the importance of historical perspective given changes in construction of fields and the impact of new developments that are both internal and external to the academy.

The field emerged during the 1930s and 1940s in English and history departments and, to a lesser extent, art history. When founded in 1951 the American Studies Association (ASA) adopted the goal of studying the country's culture and history as a whole. The first major paradigm—the myth-and-symbol school—interpreted stories and images as reflections of ideas that signified a unified national character. Comparable to modernization in area studies, however, the initial paradigm lost credibility. From the mid-1960s forward, it was attacked for idealist claims. Comparable to cultural studies and Canadian studies, the field also expanded to include new interests such as urban and environmental studies as well as popular, mass, and folk culture. Along with incorporation of European cultural theory, inclusion of once neglected and marginalized subjects amplified the discourse of critique aimed at moving beyond interdisciplinary integration of existing approaches.

Broadening meaning advanced the case for definition of the field inclusive of silenced groups and regions in counter-practices grounded in pluralism and relationality rather than unity (Klein, 1996).

In 2001, editors of the *Encyclopedia of American Studies* reinforced culture as the core concept of the field (Kurian et al., 2001). Yet, Gunther Lenz proclaimed the quest for a totalizing "inter- and supradisciplinary *Einheitswissenschaft*" had been abandoned (1982, p. 60). Indicative of his assessment, as early as 1977 Campbell and Kean called an interdisciplinary notion of America a meeting place of multiple cultures and knowledges where identity, language, and space are interchanged and contested. In an updated White Paper for the ASA, George Lipsitz (2015) characterized the field as "capacious" and an "ungainly configuration." Older interests were joined by postcolonial and queer theories and more recently necropolitics, indigeneity, and disability studies. Not everyone, though, regarded ASA as a home. Patricia Limerick (1977) recalled feeling homeless as an African American scholar, even though it was a refuge for those who could not find homes in conventional organizations. Other groups, including Chicano/a scholars, also attended ASA's conferences and published in its literature. However, they too were ambivalent about belonging.

The tension between home and abroad is also apparent in German studies. The metaphor of home implies a safe space, but, Jeffrey Peck (1989) cautioned, it is predicated on vacillation between boundaries and divergent spaces. German studies is geographically constructed on the ground of a particular country, but *Germanistik* research and teaching in other countries contributes perspectives from abroad in both a geographical and an intellectual sense. Akin to American studies, the underlying definition of culture expanded beyond universalist histories and canons of high art to include social and anthropological meaning, low and popular art, and scholarship on film, feminism, and ethnicity, thereby resituating research and education beyond nationhood while contributing to a growing trend toward transnationalism in scholarship. Peck called the field a discursive formation or strategic location for reflecting on movements across literary, political, and sociological discourses. He suggested feminism offers a model of positioning along multiple axes and contradictory discourses and practices, at the same time it interacts with disciplines and other fields. Both have implications for identity.

Identities

Identity is not exclusive. Individuals who affiliate with American studies, for example, also see themselves as literary critics, historians, art historians, musicologists, and film and media critics. The same may be said of individuals in other fields. Moreover, fields have plural identities, ranging from tight bounding of disciplinary objects to external influences to formation of new domains and intersections with other fields. Citing sustainability as a model, Fernandes and Phillippi (2017) recalled its dual origin as both a social paradigm and a field involving areas of science related to social, political, and economic problems of management. "Creolized disciplines," Cornwall and Stoddard observed, can be emergent, alternative, and resistant (1999, p. 162). W. J. T. Mitchell's (2002) account of visual studies captures challenges for representing any field. He argued a distinct object is crucial to achieving status as a discipline. However, visual studies has ties with other areas, including art history and aesthetics, semiotics, film and television, digital media, history of science, optical imaging, and philosophical inquiries.

Even with crossfertilization and overlaps, however, the boundary work of labeling continues. Naming, Armstrong and Fontaine (1989) explained, entails gathering and sorting, comparing and contrasting within an evolving sense of reality. It demarcates territorial relationships in Mike Rose's notion of political-semantic webs. Kenneth Burke's (1996) concept of terministic screens highlighted the role of nomenclature in filtering, directing, and redirecting attention. It is not simply a reflection of reality. It is also a selection and a deflection, including the space of institutional location. In the field of composition studies, to cite another example, identities differ depending on whether the site of writing support is a departmental lab or a center; its personnel are classified as staff, tutors, or consultants; and coursework is a generic skills class, technical writing, or writing-across-the-curriculum. At the same time, efforts crossfertilize. Derek Mueller (2017) retained the term *discipline* for writing studies but depicted it as an interconnected network spanning locations while generating new forms of knowledge, insights, and questions. Media and communication studies further illustrates a network conception of a field.

In a state-of-the-art account Briggle and Christians (2017) reinforced the boundary work of naming. They found titles as varied as communications, communication studies, and communication science as well as rhetorical studies, media studies, mass communication, and media ecology. To the extent studies of communication and media remain elements of existing disciplines, Briggle and Christians explained, a separate identity does not emerge. This process is not unique to media and communication studies. Miller (1982) also cited cross-cutting principles and shared components in his typology as types of interdisciplinarity. Cross-cutting principles may become so assimilated into the repertoire of a discipline they may not be recognized as foreign. Examples include concepts of role, exchange, and social processes such as acculturation, socialization, and modernization. Shared components include methods such as statistics as well as survey and interview techniques.

Disciplines were initially dominant locations of media and communication studies. Prior to 1990 communication studies at Columbia University were under the umbrella of sociology until a degree-granting graduate program in communications was created. When the University of Illinois at Urbana-Champaign established the first doctoral program in communications in 1947, all faculty held PhDs in established disciplines. Today, though, most faculty at degree-granting programs have PhDs from communication programs. Holley (2017) made a similar observation about women's/gender studies, environmental studies, race/ethnic studies, and human ecology. Yet, media and communication studies straddles the goals of serving mass media in professional training and critiquing it. New technologies and a trend toward big science also foster a transdisciplinary universalism that prioritizes cognitive and natural sciences over social scientific and humanistic approaches. Describing the kindred field of mass communications, David Scholle (1995) called it a boundary discipline situated between liberal arts and professional schools. It was constructed as a practical enterprise in schools of mass communications and speech departments. Yet, invoking G. J. Shepherd's roster of descriptors, Scholle explained an undisciplinary construction of the field would continue to service disciplines. In contrast, an antidisciplinary stance would reject foundations in the mode of critique. A radical interdisciplinary construction, in turn, would remain a locus of experiment and action, without settling on a single formation and incorporating critique of older definitions and practices.

Different constructions also appear in a final example of identity. Akin to other fields, studies of childhood and children's rights hinge in no small part on disciplinary affiliation. Moody and Darbellay (2018) distinguished childhood studies, which consider sociology and anthropology core disciplines, from children's rights studies, which consider legal studies, psychology, and sociology core domains of expertise. Both fields assert multi- and/or inter-disciplinary approaches. Yet choices of methods, research objects, and publication strategies differ, including representation of seminal work in state-of-the-art handbooks. Even so, commonalities appear. The New Childhood paradigm, for instance, is defined by five cross-cutting characteristics: study of normal childhood; critique of the conventional perspective of socialization to distinguish the lifeworlds of children; children's agency and voice; structural constraints on childhood, factoring in contexts such as economics, technology, and urbanization; and use of ordinary social scientific methods to study children and childhood. Even when acknowledging points of interactions, though, specialists address them in different ways.

Mindful of both differences and overlaps, Moody and Darbellay proposed the concept of interfield interdisciplinarity. Interfield interdisciplinarity, Moody and Darbellay emphasize, is a macro form of interdisciplinarity. It neither merges nor removes boundaries. It identifies them from the outset, rendering differences explicit but not at the expense of shared concepts, in this case including agency and federated concepts such as well-being. Interdisciplinarity is thus configured as a gradient of crossdisciplinary forms, not absolute states. An interfield conception also acknowledges both multidisciplinary juxtaposition of separate perspectives as well as interdisciplinary movement past dichotomies, binaries, dualisms, and analytic reductions. Identifying points of both divergence and convergence leads to fruitful dialogue about shared definitions, federating concepts, and a possible theoretical framework. Moreover, Moody and Darbellay stress, dialogue across differences is not simply an additive accumulation of theories, concepts, or methods. It is the starting point for integration that fosters vertical disciplinary deepening while being open to horizontal interdisciplinary questioning between disciplines.

Contexts and Intersections

To reiterate, context matters. In comparing women's studies in the US and India, Ellen Messer-Davidow (2017) demonstrated why a field, like the old adage about real estate, depends on location, location, location. Women's studies emerged in the US and India during the same decades from socio-political activism. However, it was institutionalized, produced, and circulated differently in the two countries. In the US academic legitimation required complying with performance norms for research, teaching, and service, distancing it from extra-academic activities. Many scholars also rely on familiar disciplinary methods and topics. With the exception of websites, publishing does not span sectors. In India fewer academic units exist. However, the field's institutional forms are more integrated across sectors while hybridizing academic and Indigenous knowledges in social-change projects. Thus it spans academic centers, associations, unions, NGOs, official government agencies, and grassroots women's communities. And, circulation of discourse depends on in-person modalities to reach illiterate women. Dispersion gives the field greater reach, but not free of pressures in particular sectors.

Spatial metaphors reveal the complex relationship with disciplinarity. Early in the history of the field, Annette Kolodny (1980) described literary criticism as a minefield. She recalled feminist critics had the choice of being camped on the edge or defusing it to avoid being marginalized as a tributary in the backwater or confined to a ghetto. The response was not a strict matter of either/or, however, rather a "both-and" model; feminist boundary work operates simultaneously within disciplines and in opposition to them. In order to change disciplines, one program coordinator remarked, women's studies had to be "of them, in them, and about them" (cited in Boxer, 1982, pp. 671, 693). Addelson and Potter (1991) observed the field offers intellectual community and an institutional site for feminists who do most of their work in disciplines, legitimating gender as a category of analysis. Within disciplinary locations, Hartman and Messer-Davidow (1991) likewise noted, they disperse centripetally into specializations. Within the shared space of women's studies, they move centrifugally to "cross-disciplinary" research and teaching. To no surprise, then, Catherine Stimpson described women's studies as a "portfolio of maps." Implications follow for identity (1992, p. 251) .

At the macro-level, Steen Wackerhausen (2009) explained, professional identity is a public face reinforced by shared behaviors and modes of practice. At the micro-level it is particularized. The latter form of identity occurs in three ways: through self-descriptions such as a sociologist, a historian, or a biologist; in positions such as a director, a faculty member, or a student; and in disciplinary locations. In the third case, Shira Wolosky (2013) emphasized most feminist work was still occurring within disciplines. Comparably Robyn Wiegman (2002) suggested women's studies is both disciplinary and interdisciplinary. And, a year before Coyner (1991), Diane Elam (1990) defined it as both a discipline of difference and an interdisciplinary discipline. Elam contended departmental location does not deprive members of radical politics. Instead, it harnesses funds and tenure lines, drawing strength from disciplines without being reduced to them. Robyn Warhol (2002), in turn, urged paying attention to the changing character of disciplines. Interdisciplinary scholarship and teaching are expanded at more sites.

Susan Stanford Friedman's (2001) description of her work illustrates what dual interests mean for individuals. She called literary studies her intellectual home base. It offers a platform of knowledge about literature, narration and figuration, representation, and a methodological base for reading texts. Her political home base of feminism raises questions about gender, power relations, other systems of stratification, as well as an ethical commitment to social justice. From these two homes, Friedman journeys to other interdisciplinary homes as well. Travel stimulates new ways of thinking, dislodges unquestioned assumptions, and produces new insights and questions. Her metaphor of travel also exposes the intersectionality of interests. Kimberlé

Crenshaw (1989) coined the term *intersectionality* in a legal context to signify interaction of forms of discrimination based on race, sex, and class. It can also be a metaphor for intersections of interests stimulated by travel. When accounting for feminist and gender studies in the Modern Language Association's 1992 *Introduction to Scholarship in Modern Languages and Literatures*, Naomi Schor recalled the 1980 edition lacked an essay on feminist criticism. Moreover, gender studies was presumed to encompass feminist criticism and theory as well as studies of masculinity, sexuality, and lesbian and gay perspectives. Description has changed, however.

In the updated *Introduction,* Donadey and Lionnet's (2007) account of "feminisms, gender, sexualities" portrayed boundaries of the fields as more porous today. They reported intersections with race and ethnicity; migrations, diasporas, and borders; and cultural studies. American Indian studies also exhibits hybridity with Indigenous studies and the discourse of decolonization. Both hybridity and intersectionality may be likened to Lionnet and Shih's (2011) concept of creolization of theory. It is rooted in specific historical instances and places at the same time distinctions of sex and gender combine with a poststructuralist anti-essentialism that evolved into gender studies and underwrites lesbigay studies. Postcolonial studies draws on Marxist critiques of power and imperialism. Moreover, postcolonial criticism includes marginalized literature while critiquing the ideology of colonial domination and Orientalist thought. Thus the field is both activist and adversarial. Postcolonial and other studies shaped by the discourse of critique are also democratizing humanities-based disciplines and fields by moving beyond older paradigms of analysis. The scope of disciplines has expanded as a result.

Friedman's chapter, "Migrations, Diasporas, and Borders," in the 2007 updated edition of the *Introduction* reported concepts of mobility and exchange became of interest in literary studies during the 1980s. They emerged from postcolonial studies, travel writing, and interdisciplinary fields centered on identity, including ethnic, gender, and sexuality studies. Widened impact on the discipline dates from the return to history and the rise of interdisciplinary cultural studies during the early 1990s. Expanded interests do not necessarily or strictly cohere into a single school of theory, methodology, or institutional form. They are interwoven and mutually constitutive. Transnationalism and globalization have also become prominent, pushing beyond "interdisciplinary" integrations of existing approaches to a more "transdisciplinary" scope without imposing a new universalism.

In a different realm, development studies and studies of science, technology, and society (STS) intersected at the University of the Republic in Uruguay. Arocena and Sutz (2016) traced history of development studies to the Cold

War and decolonization, when the topic became part of political and academic agendas. Scholars conducted research on different aspects of development. Yet, it was considered belonging to the discipline of economics. Sustainable development, though, brings ethics into a normative approach in environments that are simultaneously natural and social. Interactions between social relations and technology, Arocena and Sutz added, consider productive forces and relations in a Marxist theory of history. Building a subfield called development, science, and technology at the University has provided opportunities for collaboration between natural and social sciences. In 1992, the University Research Council began conducting academic work intersecting with STS. Typical of many fields, discussion ensued on whether to concentrate on disciplinary agendas, such as sociology of science, economics of technology, or an integrated perspective. The latter strategy prevailed, but not without risks to individuals in the academic reward system. As a result, many young researchers defaulted to disciplinary identities. Over time, though, an STS identity developed.

In 1994 the Faculty of Sciences created a unit focused on science and development that cooperates with a Network of Development Studies. The majority in the Network are development economists and economic historians. Yet, it fostered organization of a bachelor's degree in development that, along with a master's program in economic history, offers courses related to science, technology, innovation, and development. The bachelor's degree also promoted organization of a specialized teaching module in science and technology, including a short course on interdisciplinarity, giving students tools for integration. Creation of the Espacio Interdisciplinario took a step further, influenced by the concept of a developmental university that supports integration of traditional academic functions of teaching, research, and service extension. Within this space, Centers for Climatic Change, Childhood and Poverty, and Coastal Management were also created, as well as a unit to study and to promote interdisciplinary research.

The comparative picture of interdisciplinary fields in this chapter reinforces the heterogeneity of interdisciplinarity today. Yet, echoing Brint, Turk-Bicakci, Proctor, and Murphy's survey of majors, the academy has not been transformed as a result. Angela Last reminded readers, "The reality is that the uptake of interdisciplinarity, while seemingly universally desirable, is uneven" (2018, p. 202). Updated genealogies affirm this caveat, including Craig Calhoun's (2017) account of area studies and Sheila Jasanoff's (2017) assessment of STS. Calhoun warned that without institutional conditions for reproducing a field it is "at the mercy of disciplines that may claim it, ignore it, or incorporate some ideas from interdisciplinary projects without providing ways of sustaining the intellectual conditions produced them" (2017, p. 121). In Latin American and East Asian studies, for example, a few universities established autonomous departments. More, though, set up committees or centers, leaving appointment and promotion of faculty as well as awarding PhDs to disciplinary departments. Co-optation, denial, and assimilation undermine the identity of a field and its contributions.

Jasanoff's account of STS yields lessons for other fields in an extended example. The field was slow to gain a foothold in universities, she recalled, but over roughly 50 years became respected for scholarship and other achievements. Some prominent individuals identified explicitly with STS, but the balkanization that results when resources are short led many young scholars to default to recognized disciplinary affiliations, such as anthropology, history, or sociology. Or they aligned with topical subfields in market demand, such as bioethics, environmental studies, science policy, or even nanotechnology and society. Dispersal diminished prospects for developing and anchoring definition in commonalities. Even today, the case for autonomy comes into conflict with traditional disciplines claiming work of the field, fragmenting, for example, into the history of physics or of technology, historical and sociological study of biology, sociology of science, and anthropological study of cultures of science or technology. Moreover, the field remains weakly institutionalized today in upper reaches of the academy, and it has not integrated diverse interests. Hurdles are built into both the political economy of disciplines and the field's contradictory self-understandings. The challenge of interdisciplinarity, Jasanoff concluded, is strategic positioning.

All fields, she admonished, must establish relations to their objects of study and other disciplines, to assert their boundaries and missions, and to question the premise that disciplines are coherent, unified entities. They require spaces where goals can be developed and sustained with a critical mass of economic, social, and intellectual capital, populating them with well-trained scholars, new curricula, and long-term research programs. They demand organization for survival and continuity to demarcate themselves from neighboring territories and to establish markers of originality, progress, and contributions to knowledge. And, they need to define what makes them part of a common endeavor in the larger landscape of knowledge. Despite growing attention to the field's intellectual contributions, there are few full-fledged STS departments in the United States, even fewer in Europe, and barely any in Asia or Latin America. Departments, moreover, tend to cluster in engineering schools and, with a few exceptions, have not taken hold in high-prestige research universities.

Jasanoff identified three overarching challenges STS, and by extension other fields, must overcome before taking their place as a necessary and even indispensable component of higher education: establishing credible relations with its objects of study; defining its relations to other disciplines; and asserting a stronger sense of its own boundaries and mission to avoid being co-opted and marginalized by better established, resourced, and valued domains of inquiry. Otherwise, STS and they, to echo Jasanoff, become cast as a sovereign academic currency offering nothing new without appreciation of its originality and contributions to fundamental knowledge. Reconstituting areas into a combined field of studies is often a solution to financial exigencies, whether lumping together small fields with low enrollment or bringing a field into a larger one or an established discipline. Both strategies keep them alive, but often diminished. Moreover, even under the banner of interdisciplinarity, this strategy does not ensure separate parts are anything more than a shotgun marriage. Jasanoff cited two types of cartographic metaphors underlying the idea of interdisciplinarity. In one configuration, disciplines are lined up tightly, as in a map of the contiguous United States, with shared boundaries and no gaps between them. In contrast, an archipelago akin to Indonesia depicts them as idiosyncratically bounded islands or interdisciplines with unexplored waters between them.

STS, Jasanoff concluded, resembles an archipelago charting new territories in seas of the unknown. On the first map, a new interdiscipline comes into being principally through exchanges among scholars belonging to one or another established disciplinary community and trained in its forms of reasoning and research practices. On the second, an interdiscipline is an autonomous formation situated among other disciplines. The future of STS, and once again other fields, by extension, depends on redrawing the map of the disciplines to demonstrate they are all islands of happenstance, with unmapped waters between them. Then they can claim a space for themselves as another fertile territory in these wide waters, offering resources for understanding some of humanity's most impressive accomplishments, but without threatening anything achieved, or yet to be achieved, in other quarters of the disciplinary archipelago.

In yet another extended example, Mindy Fenske (2007) called particular attention to the challenge of negotiating new territories when multiple audiences are involved. Fenske used a review of three anthologies as an occasion for an extended example of defining and practicing a field using performance studies and its interdisciplinary contours:

- Philip Auslander's (2003) *Performance: Critical Concepts in Literary and Cultural Studies*

- Judith Hamera's (2006) *Opening Acts: Performance in/as Communication and Cultural Studies*
- D. Soyini Madison and Judith Hamera's (2006) *The Sage Handbook of Performance Studies.*

Fenske herself experienced vagaries of identity after being hired in 2001 into a performance studies line within a speech communication program located in a Department of Theater, Speech, and Dance at the University of South Carolina. In 2002, the speech communication program was renamed a Department of Theater and Dance and relocated to a Department of English Language and Literatures.

However, Fenske was not familiar with the culture of English departments and, vice versa, faculty in English knew little about communication studies. Yet, she saw relocation as an opportunity to embody the interdisciplinary identity of performance studies by bridging English and communication studies with relationships of performance, literary, rhetorical, linguistic, and film and media studies housed in or affiliated with the Department. Doing so required reshaping her own scholarship to speak to multiple audiences, including visual studies and rhetoric in communication and composition studies. Fenske's experience is not unique. Many scholars and artists, she reported, find creative ways of sustaining a performance identity while teaching in rhetorical studies, literary studies, theater history, cultural studies, women's studies, African American studies, intercultural communication, visual studies, and communication pedagogy.

Shannon Jackson's (2004) disciplinary genealogy, *Professing Performance,* revealed epistemological and institutional disagreements among individuals who presumed to share a specialty of performance at the same time they were negotiating the form and the content of interdisciplinary work to render it recognizable by multiple audiences. Differences in genres of professional communication complicate the challenge. In this field they include evaluating performance as scholarship versus a traditional monograph, relying on social scientists' criteria of objectivity and replication in laboratory work, and making the case for community-based and local performances to administrators who are accustomed to national and international forums. The challenge of negotiation is further complicated by the contested nature of both performance and interdisciplinarity. Fenske identified three themes that appear in other accounts. In the first, scholars borrow a concept or method from another field. In the second, theoretical collage or montage perpetuates the pejorative reputation of being a jack-of-all-trades and master-of-none. Jackson's concept of synechdochic fallacy occurs in the third theme, presuming one approach, movement, or piece of scholarship can represent entirety of a discipline. Definitions of fields also differ.

In introducing the four-volume *Performance* collection Auslander (2003) presented a seven-part dictionary definition marking presence of the keyword in a host of areas. His roster included theater, performance, dance, and film studies as well as the profession of law and disciplines of anthropology, sociology, art history, philosophy, literary criticism, and speech and oral interpretation. Fenske also noted other subjects and orientations in the collection: media theory, communications, cultural studies, feminist theory, ecology, Marxism, structuralism, poststructuralism, and postmodernism. This expanse reinforces an overarching lesson about fields. The objects and subjects they develop are co-located, as the foundation for a separate domain and subinterests within traditional disciplines. The remaining three volumes of the collection are subdivided into topic areas based upon conceptual issues (volume 2 covering representation, textuality, bodies, audiences, and spectatorship), relations between disciplines (volume 3 including sciences and social sciences), and sites or locations (volume 4 accounting for identity, visual and performance art, and media and technology).

Auslander's opening description, though, did not mirror those categories. He began with Richard Schechner's distinction between performance research on objects of criticism that are aesthetic performances and critical analysis. He then previewed the variety of performing art objects across the four volumes: genre, performer, audience, and cultural performance. And, he cited models for understanding social aspects of performance, though Auslander continued to highlight criticism and its leading thinkers. For their part, Madison and Hamera also began *The Sage Handbook of Performance Studies* with a different strategy of representation, citing three founding scholars and situating performance as an interdiscipline from a communication-oriented perspective shaped by oral interpretation and literary performance. Rather than approaching definition through objects, they began with the process of elocution, aligning the field with the often elided history of oratory and speech. Here too, though, the *Handbook* is not organized along lines of the introduction. Sections are divided into relationships of performance and theory, history, narrative and literature, pedagogy, ethnography, and performance and politics. This organization accentuates performance as action and aesthetic. Individual essays, Fenske added, reflect tensions between process orientation and the tendency of handbooks to bound representation in categories, definitions, and conceptual surveys.

In comparison, Fenske depicted Hamera's separate collection, *Opening Acts*, as an exemplary enactment of interdisciplinarity that depicts the complex terrain of performance scholarship. This volume treated performance research and method as critical communication study, though pinned definition to specific methodological and theoretical descriptions. Thus it eschewed confining performance scholarship to tidy categories while minimizing diverse voices. Scholarship is instead an amalgamation. Reflecting comparatively on the three collections, Fenske concluded they demonstrate interactions between ideas and concepts from multiple disciplines, with potential for combining them synergistically into something new. Each anthology and most of their essays, she added, illustrate the type of interdisciplinary collaboration that combines emphasis on collaboration in science and solo authorship in humanities. They also enact Jackson's notion of the shifting terrain of interdisciplinarity, acknowledging variation and heterogeneity while recognizing that contingency and change produce alternative histories and the possibility of different futures. These and other examples bring discussion back to the nature of interdisciplinary fields.

McKeon's conception of interdisciplinarity as an architectonic art underscores the boundary work entailed in constructing and maintaining them. Friedman's notion of productive travel is an index of their dynamic interplay with fixed institutional structures. As a result, an individual might be doing feminist work in some way without being a formal member of the Women's Studies Association. A historian or an anthropologist might be engaged in study of a particular world region without identifying as a specialist in area studies. A chemist or a biologist might be investigating consequences of new discoveries or innovations without being a member of a STS program. Students may not affiliate with a formal studies program, either. For example, some are earning degrees in philosophy, sociology, and economics but pursue dissertation topics focused on media and communication.

Roland Barthes' (1997) proclamation that interdisciplinarity consisted of creating an object which belongs to no one is a popular characterization. However, it sidesteps the fact objects are always claimed in some way. An object might also be an assimilated component of disciplinary practice, part of the repertoire of an interdisciplinary enclave, or central to a transdisciplinary conception of a field. Intersectionality heightens awareness of multiple locations as well. The transdisciplinary operation of cultural studies, Douglas Kellner (1995) contended, draws on several fields to theorize complexity and contradictions of media, culture, and communications while pushing boundaries of class, gender, race, and ethnicity. Dölling and Hark (2000) associated TD in women's and gender studies with critical evaluation of terms, concepts, and methods that transgress disciplinary boundaries. Vickers (1997) linked trans- and anti-disciplinarity with movements that reject disciplinarity in whole or in part while raising questions of socio-political justice. And, in family science Libby Balter Blume (2014) advocated moving past both multidisciplinary juxtaposition of disciplinary and theories as well as interdisciplinary borrowings and integrations that do not disrupt the status quo. A transdisciplinary identity encompasses both an overarching framework and complex problem solving involving stakeholders in what Blume calls a form of border work.

Ultimately this chapter and indeed the entire book raise the question of whether there is a distinctive interdisciplinary logic, regardless of field. When claiming better and more accurate or complete knowledge, Elizabeth Bird (2001) contended, interdisciplinarity follows the same logic as disciplinarity. Other logics, though, shape fields: including plurality, combinations, and transgressions. Jasanoff framed the difference as an internal desire for a high-church disciplining and a broad-church approach to keeping it open. Scott Frickel (2004) contended the key boundary problem lies in perforating existing boundaries and/or inventing porous ones. Disciplinary logic would construct power relationships that subordinate alternative claims and impede boundary crossing. In contrast, interdisciplines would encourage permeability and shifting between logics in different levels, arguments, or stages in the flow of ideas, people, skills, practices, and resources. Interdisciplines, he concluded, are hybridized knowledge fields located between and within existing disciplines. As a result they interact with other fields, and they exhibit more epistemological variability than disciplines. Comparably, Mathias Friman (2010) stated disciplinary logic promotes monopolization, while interdisciplinary logic promotes plurality and by extension permeation. Part II will continue to examine related tensions, though turning to deeper analysis of Kinchy and Kleinman's (2003) category of strategic boundary work that advances targeted stakes.

PART II

DYNAMICS OF PRACTICE

4
Communicating and Collaborating

> The structure of organizational boundaries consists of *mediating* objects, the *coupling* that ties people to them, *properties* that enable and constrain *interaction* and the *identity* of both individuals and groups. Social practices include *objectification*—the process by which objects are (re)created; *identification*—the process by which people couple with *objects*; *interaction*—the way in which people act on each other; and *emergence*—the complex process by which higher-order *properties* of boundaries are created. All of these elements of boundary are integrated by a basic social process called *boundary weaving* in which people *navigate* the elements of boundary structure at the same time they *(re)create* them.
>
> **Timothy Hans Kannegieter (2010, p. 3)**

Communication and collaboration go hand in hand. Some areas are collaborative by nature. The National Academies of Science, Engineering, and Medicine's report on team science noted sizable groups in physics and genomics (Hilton and Cooke, 2015), and Cummings and Keisler (2005) cited oceanography and cognitive science. Groups differ, though, in size, goals, mix of expertise, and degrees of coordination and integration. Moreover, the NASEM report noted, while team members work in academic settings they also affiliate with industrial and federal laboratories as well as research facilities such as observatories and particle accelerators. No single model prevails. Introducing a book on communication and collaboration in interdisciplinary research, Crowley, Eigenbrode, O'Rourke, and Wulfhorst (2014) highlighted affective, organizational, and instrumental dimensions along with linguistic, relational, cultural, and socio-political variables. Introducing another volume on communication in teams, Klein, Misra, and Lotrecchiano (2020) reported the topic entails both pragmatic and conceptual dimensions while spanning intrapersonal, interpersonal, organizational, institutional, and socio-political levels.

This chapter explores dynamics of boundary weaving Kannegieter identified in the opening epigraph. In doing so it joins Lury et al. (2018) in treating

Beyond Interdisciplinarity. Julie Thompson Klein, Oxford University Press. © Oxford University Press 2021.
DOI: 10.1093/oso/9780197571149.003.0005

interdisciplinarity as a verb rather than a noun. The difference is not trivial. It accentuates process over product and final outcomes. In the same volume, Tahani Nadim (2018) highlighted crafting over recovering and inventing over finding. The chapter begins by defining the nature of talk across boundaries, including pidgin and creole forms of language, linguistic and social dynamics of communication, a culture that fosters them, epistemic dimensions of dialogue, and relational thinking. It then focuses on collective identity in teams and stages of collaboration, followed by a section on the nature of integration and differing assessments of its centrality to crossdisciplinary work. The chapter turns next to public engagement and community-based research, in the process moving beyond narrow characterization of translation as application and transfer to highlight intersubjectivity, communicative action, and participatory research. It concludes by illustrating translation boundary work in two cross-sector case studies, an urban planning project and a waste management project involving both academics and community stakeholders.

Talking across Boundaries

Crowley, Eigenbrode, O'Rourke, and Wulfhorst's (2014) introduction to their edited volume on communication and collaboration in crossdisciplinary research noted the paucity of books dedicated to the topic. Yet, the challenges are widely recognized, especially for crossdisciplinary work, whether the number of participants in joint efforts are 2, 20, 200, or even up to 2,000 in large and distributed projects and research programs. The communicative actions of collaborators are a form of boundary work that requires the ability to talk across boundaries. In a book devoted to the topic, Burtis and Turman (2006) contended all group communication involves boundary spanning. However, spanning, transferring information, sharing meaning, and negotiating take on greater importance in crossdisciplinary collaboration because it requires bridging knowledge domains with different logics, methods, and languages. A state-of-the-art report on interdisciplinarity even asserted communication lies at the heart of interdisciplinarity (NASEM, 2005), while Jacobs and Frickel (2009) defined interdisciplinarity in essence as communication and collaboration. Klein, Misra, and Lotrecchiano (2020) cautioned, though, the underlying concept of communication is ambiguous. Academic studies alone range from conversation analysis and small-group dynamics to complexities of dialogue in social networks and professional education. Furthermore, communication is not just a linguistic phenomenon. Joann Keyton (1999) called it a social process.

Based on studies of interdisciplinary research teams, Jessica Leigh Thompson placed collective communication competence at the heart of collaboration. It is neither simple nor formulaic. Rather, it is dependent on "numerous interrelationships among communicators, contexts, goals, and the participants' abilities to simultaneously be appropriate and effective" (2009, p. 281). Yet, Chou and Wong (2015) warned, connections among disciplines are not always clear to team members. The question of whether communication is even possible also remains a matter of debate. It hinges on whether domains of knowledge are commensurable. Britt Holbrook (2012) identified three major positions. The first, and dominant, view asserts translation across disciplinary frameworks is possible. Integrating separate languages can generate a new common understanding through reciprocal comprehension and consensus, bringing about mutuality and intersubjectivity. The second view considers translation impossible because disciplinary languages are in principle and often in fact incommensurable. Communication, then, can only occur by adopting the language of another discipline. The third view advocates creation of a new language.

Debra Journet's (1993) study of Smith Ely Jelliffe's effort to advance a theory of psychosomatic medicine illustrates the first view. In attempting to connect neurology and Freudian psychoanalysis, Jelliffe translated values and principles of one field into the language and discourse of another. His attempt, however, was criticized for being too general or philosophical, while exposing conflicts between quantitative and qualitative approaches. He then shifted strategy by creating intertextual links between internal medicine, neurology, and psychoanalysis. In doing so, Journet suggested, Jelliffe in effect narrativized neurology and scientized psychoanalysis, casting each in the other's image to suggest compatible evidence and interpretations. He inserted empirical data into psychoanalytical case history while explaining psychoanalytic concepts such as conversion hysteria in quantitative, experimental language. Jelliffe's effort also illustrates the role of interlanguage, which Holbrook and Klein (2013) likened to pidgin and creole forms of communication. Bilingualism is a popular metaphor of interdisciplinarity. However, mastery of two disciplinary languages rarely occurs.

To recall Chapter 1, Peter Galison (1997) described pidgin and creole as contact languages. In linguistics pidgin is also a technical term for an interim tongue, while creole is a new first language. Drawing on experience in bioethics, Maurice DeWachter (1982) observed that ideal models start with the assumption individuals will suspend their disciplinary/professional worldviews. Yet, he advised, the best chance of success lies in starting by translating a global question into the language of each discipline, then working back and forth iteratively.

By constantly checking the relevance of each answer to a core question, no single answer is privileged. In an early study of interdisciplinary discussions, Gerhard Frey (1973) reported they usually take place on a level similar to a popular scientific presentation. They become more precise as individuals combine everyday and specialist languages. However, based on analysis of interdisciplinary conversations in six faculty seminars at three universities, Myra Strober (2011) reported persistent differences in languages and styles emanate from entrenched disciplinary cultures and habits of mind. Bridging them requires creating Rick Carpenter's (2009) notion of a communication boundary space.

Others concur. Wilhelm Vosskamp (1994) admonished the quality of outcomes cannot be separated from a shared language culture, while Schmithals and Berhenhage (2004) deemed a cooperation and communication culture essential for trans-disciplinary projects. They described it as a targeted approach that identifies interfaces where work of one participant is necessary for another. Michael Gibbons (2008) further argued the metaphor of transaction space shifts from translation to dialogue. Given the entwined relationship of language and conceptualization, Mathias Friman (2010) proposed discourse theory is an appropriate methodology for studying communication across boundaries, Choi and Richards' (2017) discourse analysis of transcripts from meetings of eight groups suggested collaborators occupy epistemic domains individuated by what they know and the epistemic stance by which they situate themselves. Epistemic rights and status flow to individuals whose knowledge is deemed most relevant. At Stage 1, individuals advance territorial claims by displaying, deploying, and directing their knowledge; making connections; and clarifying terminology. Stage 2 involves coordination, and at Stage 3 affiliative talk facilitates integrated teamwork.

In a separate study, Osbeck and Nersessian (2010) drew insights from three configurations in two research labs, one in tissue engineering and the other neural engineering:

1. Explicit references to professional or disciplinary affiliations;
2. Cognitive activities and knowledge construction;
3. Relationships with artifacts and objects, such as anthropomorphizing cells by calling them "guys" and referencing their feeling effects.

In the first configuration, individuals positioned themselves through disciplinary anchoring. In the second they moved to relational thinking while aligning with methods or ideas of other individuals, labs, and research traditions. In some cases, though, distancing occurred as individuals made different choices in technical texts and levels of organization.

Reflecting on their findings, Osbeck and Nersessian described a laboratory as a local discursive community constituted by rules, conventions of conduct, and terms and concepts established through interactions. At the next level the overarching storyline of a unit is interwoven with smaller storylines representing particular projects and interactions, then at a broader level each laboratory is embedded in wider communities of practice such as disciplines and fields. Identities differ as well. In the first configuration individuals tended to associate with either biology or engineering. Some, though, identified with a hybrid of biomedical engineering or bioengineering. Negotiations in an interdisciplinary culture, Osbeck and Nersessian concluded, occur as researchers find a place fitting their backgrounds and skills. The process is not always clear-cut, however, especially in cutting-edge fields. Definition of what it means to be a hybrid biological engineer, for example, is not so well established as being an engineer or a biologist.

Based on a scoping study to cross boundaries of human and physical geography, Bracken and Oughton (2006) identified three linguistic practices that appear in other studies as well. The first practice—*Dialects*—marks the difference between everyday and specialist languages. It requires translating their meanings in order to arrive at a common interlanguage. The second—*Metaphor*—develops thinking in new directions. The third—*Articulation*—deconstructs disciplinary knowledge. Meanings in their original contexts still differ, but one language does not dominate. Physical geographers associated the word *dynamic*, for example, with stream discharge that varies by moisture conditions in short timescales. In contrast, social scientists associated it with rapid changes over longer periods. In another instance, when talking about her work the physical scientist defined *catchment* as an area of land defined by drainage boundaries of a river. Social scientists, though, felt the term had little meaning for social and economic processes related to landscape. The physical scientist also acknowledged human activities affect the environment but discounted their relevance to physical processes. By working through differences, participants paved the way to Bracken and Oughton's notion of interdisciplinary academic capital for collective identity deemed crucial to arriving at more than a multidisciplinary assemblage.

Collective Identity

Young teams, Anthony Stone (1969) reported, tend to exhibit secondary-group relations characterized by thinking in terms of "I." In contrast, primary-group relations are characterized by thinking in terms of "we."

Violina Ratcheva's (2009) concept of relational capital highlighted the importance of trust, reciprocity, and strong ties in arriving at "we." The importance of relationality calls to mind Lang et al.'s (2012) conception of trans-disciplinary research as an interface practice. Building capital at an interface requires Lotrecchiano and Misra's (2020) notion of cognitive shifts at both individual and organizational levels. The shifts foster collective thinking or sense-making through open, deep, and continuous interactions and reflections that foster both self- and collective efficacy. Freeth, Clarke, and Fam (2019) cautioned, however, that working at the interface is not a strictly linear progression from Stone's "I" to "we." They based their caveat on a trans-disciplinary sustainability team at a German university. Its heterogeneity increased as it moved from 8 principal investigators to 23 members, including partners outside the academy. Members were also located in institutes with different methodologies, disciplinary expertise, ways of interacting, and reward structures. Diversity, though, complicates communication.

Tensions surfaced in a series of retreats. During the first one, discrepancies between PhD students' individual research proposals and the overall agenda were resolved by letting them prioritize their needs. That decision, though, meant foregoing participation in selecting case studies, integrating work packages, and realizing overall goals. At a second retreat a year later, familiar tensions arose between participants favoring an emergent approach to the process of research integration and those preferring a tightly planned and managed approach prioritizing directed outcomes. At a third retreat tensions surfaced around differing interpretations of key concepts, though boundary objects of transformation, systems thinking, and collective and transdisciplinary research practices facilitated collective coherence. Based on their experience, Freeth, Clarke, and Fam emphasized the importance of moving from serial monologue to engaged dialogue. It does not mean surrendering individual agency. They advised the "I" needs to remain in a healthy tension with the "we" of a team.

Comparable to Freeth, Clarke, and Fam, as well, Bennett, Gadlin, and Marchand (2018) urged dialogue rather than competitive debate in the US-based National Institutes of Health guidebook to *Collaboration and Team Science*. Dialogue is still slowed or even thwarted, however, by Evelyn Brister's (2016) concept of disciplinary capture. Capture occurs when one discipline in an interdisciplinary project takes precedence over others, skewing standards, values, and methods. It also imposes one disciplinary language and worldview, impeding communication among collaborators and generation of research questions and designs that differ from epistemological frameworks of the contributing disciplines. Bethany Laursen's (2018) notion of collaborative interdisciplinary reasoning defines the capacity required for productive dialogue around a common goal. By co-applying, co-revising, and co-creating intersubjective standards, participants generate shared assumptions about what counts as "good" reasons and inferences throughout all phases of a project. The prefix "*co-*" appears frequently in literature on crossdisciplinary and cross-sector work, but Gunawardena, Weber, and Agosto (2010) distinguished important differences.

Gunawardena, Weber, and Agosto's scheme took the form of a stage model:

1. The first stage—Coordination—entails exchange of information, mutually beneficial activities, common purpose, informality and low risk, retention of authority in each organization, and separate resources and rewards without needing a common mission, structure, or planning.
2. The second stage—Cooperation—involves exchange of information, mutually beneficial activities, and common purpose. It adds shared resources, greater formality, a level of risk, a degree of planning and dividing roles, compatible missions, communication channels, mutually available resources, and rewards.
3. The third stage—Collaboration—entails exchange, activities, and common purpose as well as enhancement of others' capacity when sharing resources, a durable and pervasive relationship, increased risk, creation of a new structure, commitment to a common mission, comprehensive planning and clear communication channels at multiple levels, pooled resources, and determination of authority in a collaborative structure.

Gunawardena, Weber, and Agosto's definition of the third stage drew insights from multiple literatures: including organizational science (concerted action with shared rules, norms, and structures), library and information sciences (a well-defined relationship with shared meaning and goals), and education (shared responsibility and authority for planning and integration). Like Freeth, Clarke, and Fam they contended collaboration does not necessarily require surrendering individual positions. Individuals may still retain strong disciplinary commitments, identities, and repertoires, but they subordinate their priority to a collective goal. Based on an ethnographic study of a cancer research center, Centellas, Smardon, and Fifield (2014) proposed calibration of viewpoints does not inevitably signal or cause blurring or softening disciplinary boundaries. It unfolds in daily negotiation and alignment of personal identities, commitments to disciplines, and customs of research groups. An extended case study presents a fuller picture of the process in a stage model that moves from *Positioning* to *Purposing*.

Piso, O'Rourke, and Weathers (2016) analyzed negotiations in a workshop on the ontology of fog systems and methodologies. Coastal fog is a source of water and nutrients in ecosystems of forests along Pacific coasts of North and South America. It can have negative impacts, including introducing pollutants from the atmosphere. Research has tended to concentrate in particular domains, though atmospheric chemists and oceanographers have collaborated at the marine boundary. Conversations classified as *Bounding* and *Layering* focused initial attention on particular issues or problems, while conversations around *Purposing* and *Modeling* entailed evaluating feasibility of conditions.

1. During the first stage—Positioning—individuals located themselves within group conversations. Short introductions revealed limits to both their own and others' expertise, a crucial starting point in forging collective identity and a collaborative community.
2. In the second stage—Bounding—three patterns of conversation facilitated demarcations: distinguishing entities within versus outside fog as a system, differentiating entities within it from one another, and organizing within it to suggest interfaces for integration.
3. At the third stage—Layering—two patterns of conversation appeared: locating one subsystem within another on broader spatial or temporal scales and relating entities or processes from one subsystem to entities and processes constituting another subsystem.
4. The fourth stage—Focusing—simplified complexity by reducing the number of variables. The amount of possible connections can be overwhelming, diminishing an integrated approach. Focusing establishes new boundaries that determine which are most relevant.
5. In the fifth stage—Purposing—conversations focused on projects that could connect social-ecological impacts across subsystems. It also moved through layers of the system such as fog oceanography and atmospheric chemistry.

The five stages are not absolute, however. Individuals continue *Positioning* throughout other stages while evolving from the singularity of "I" to a collective "we." Dialogue facilitates *Bounding* around a construct that does not erase differences but contextualizes them. *Layering* fosters relationality, and *Focusing* establishes boundaries around a project that *Purposing* then refines. In a separate study, Hall and O'Rourke (2014) identified optional approaches at designated phases. When launching a project, participants can use a variety of tools to externalize individuals' values, assumptions, and knowledge. When moving to integration, participatory modeling may range from simple cognitive mapping exercises to complex models of system dynamics. Mapping also helps visualize spatial aspects of a project, including local stakeholder knowledge and project design. When generating alternatives, scenarios focus on solution-oriented knowledge and, in conjunction with co-constructing a systems model and visualizing informational tools, facilitate moving forward. At the final stage Delphi method can help weigh alternatives. No single best approach dominates, then. Participants need to be familiar with a repertoire of strategies and mechanisms, while heeding demands of context.

Integration

As the foregoing examples suggest, integration is also a key factor. Klein and Newell called integration the "acid test" of interdisciplinarity (1997, p. 404), while Pohl, van Kerkhoff, Hadorn, and Bammer deemed it "the core methodology underpinning the transdisciplinary research process" (2008, p. 42). However, here too, there is no universal model or formula of integration. Zierhofer and Burger (2007) differentiated thematic, product- or problem-oriented, and social types, while Jahn, Bergmann, and Keil (2012) distinguished epistemic, cognitive, social-organizational, and communicative levels. Based on literature review O'Rourke, Crowley, and Gonnerman (2016) identified multiple means of integration: including unification by reduction, a global theory or an overarching abstract model, interconnections between fields, local theories, and micro-level integrations. In addition, they identified four fault lines of definition.

The first fault line is linear, algorithmic step models versus heuristic and constructivist frameworks. The second hinges on whether integration is cognitive in nature or factors in social and communicative aspects. The third is between integration as an individual versus a collaborative phenomenon, and the fourth is emphasis on disciplines versus societal perspectives. Illustrating the first fault line, William Newell (2001) and Allen Repko (2008) built on the step model I proposed in 1990 when advocating a 10-step progression for individual students conducting solitary interdisciplinary research projects. Deeper study of collaboration, however, revealed it was insufficient for understanding dynamics of integration, which are more complex for teams crossing disciplinary, interdisciplinary, and sector boundaries. Laursen and O'Rourke (2019) likened the socio-linguistic approach I subsequently adopted to bootstrapping in a process of iterating, patterning, and back-and-forth testing. The 25 integrative techniques in my 1990 step model are still relevant. However, they noted, some take on greater magnitude in teamwork, including role clarification, bridge concepts, common foci, and an iterative model of communicative action. Those who favor a heuristic approach, Laursen and O'Rourke added, contend integration is a creative process that lacks generalizable steps.

The nature of integration was also being questioned within disciplines. In biology, for instance, Ingo Brigandt (2010) contended integration is not a regulative ideal or an aim in itself. When explaining the evolutionary origin of novel structures, for instance, elements of disciplines could be related to handle a particular problem by linking smaller epistemic units such as concepts, methods, and explanations. Comparably, Elihu M. Gerson (2013) argued integration is typically partial rather than full epistemic and organizational unification of two specialties, an ideal often invoked but rarely if ever realized. Partial intersections form when models, descriptions, or procedures become part of the work of another discipline. Sometimes a sub-problem in one specialty, Gerson added, falls under jurisdictional expertise of another, as when the genetic problem of characterizing physical properties of genes required help of biochemistry. Observations of one specialty sometimes become a problem for another as well. The classic example of partial integration is Darden and Maull's (1977) concept of interfield theories. The chromosome theory of inheritance, for example, connected Mendelian genetics and cytology. Others have also proposed alternatives when defining the nature of interdisciplinarity.

Veronica Boix-Mansilla (2010) cited aesthetic interpretations, more comprehensive explanations, predictive models, informative contextualization, and practical problem solving. In a more radical stance based on literature review and interviews of faculty across sciences, social sciences, and humanities, Lisa Lattuca (2001) challenged the primacy of integration. She argued interaction of disciplines is a more inclusive basis than integration because it encompasses a wider continuum of forms and outcomes. Proponents of critical approaches in particular were less inclined to accept existing disciplinary boundaries than faculty with modernist and positivist views. They favored feminist, poststructuralist, and postmodernist theories. Akin to others who emphasize heterogeneity of the underlying concept, Lattuca concluded, "Different kinds of questions led to different kinds of interdisciplinarity" (p. 79).

Nonetheless, a number of traits of integration appear regardless of degree of interactions and context of both crossdisciplinary and cross-sector work. In proposing a framework for determining whether outcomes are integrative, Gabriele Bammer identified six generic questions. They encompass for what and for whom, of what, by whom, context, the how of process, and impact. In subsequently boiling them down to five questions she combined how and by whom into one question. Bammer also fleshed out determining factors: overall circumstances, a problem's history, geographical locations, and cultural differences between those affected and those charged with responding to a problem (email, 11 June 2020). Sources of authorization or legitimacy for knowledge synthesis also affect what is investigated. Some studies concentrate on organizational structure, since institutional possibilities and limits are a major determinant of the success or failure of interdisciplinary initiatives. However, literature on teamwork also recognizes cognitive, social, and psychological dynamics of collaboration.

In shedding light on the relationship of cognitive and social integration in particular, Salazar, Lant, Fiore, and Salas (2012) called attention to communication, collective understanding, and emergence in three pathways:

1. The first—social integration processes—entails common understanding of a goal in the form of shared mental models, communication practices facilitated by shared leadership, and collective understanding.
2. The second—emergent states emanating from social processes—recognizes formal interventions, norms, and technological infrastructure provide support for emergent states, and in structured interventions encourage team members to inquire about others' expertise.
3. The third—cognitive processes of knowledge consideration—emanates from social processes and emergent states, thereby facilitating assimilation and accommodation that lead to continued growth of a team's integrative capacity.

Bennett, Gadlin, and Marchand's (2018) guide, to *Collaboration and Team Science* combined cognitive and social factors in a continuum of interaction and integration, with ideal teams exhibiting the highest levels of shared leadership and decision-making. In a widely quoted early study of 24 technology assessments, Rossini and Porter (1979) identified four major types: integration by a leader or a subgroup, negotiation among experts, common-group learning, and joint integration. Reviewing them almost three and a half decades later, Andersen and Wagenknecht (2013) invoked John Hardwig's (1985) distinction between unilateral and multilateral epistemic dependence in interdisciplinary collaborations. Integration by a leader is a unilateral form of dependence while, in contrast, common-group learning is a form of multilateral dependence that derives from mutual learning and sharing the same mental models and concepts. Leadership remains a primary topic in literatures on collaboration, though Elina Mäkinen (2018) identified a shift from leader-centric models to systems thinking that recognizes complexity and relationality of social processes as well as interacting variables.

On a more granular level, Barbara Gray (2008) highlighted three sets of leadership tasks that foster integration and collaboration in transdisciplinary research teamwork. The three tasks shed further light on challenges of achieving collaboration:

- Cognitive tasks manage meaning and sense-making by creating a mental model or map of goals and methods for achieving them. Visioning occurs on two levels: content and the process of working collaboratively, while reframing requires suspending current assumptions.
- Structural tasks foster coordination and exchange through boundary spanning, conveying information across groups, brokering information flow, facilitating transactions, transferring and translating, and building linkages.
- Processual tasks, in turn, develop team dynamics by deploying interpersonal skills for joint decision-making and problem solving while exchanging information, coordinating efforts, and managing boundaries.

Leaders are often chosen on the basis of reputation in a speciality. However, tasks Gray identified affirm the need for someone who possesses not only content knowledge but also multidisciplinary breadth and familiarity with integrative and collaborative process. In examining intrapersonal qualities of leaders in fostering communication on 52 multidisciplinary medical research teams, Salazar and Lant (2018) found leaders who possess multidisciplinary breadth are best positioned to facilitate communication between team members who have never collaborated before and have little overlapping expertise. Being well versed in a variety of areas enables leaders to contribute insights that stimulate interdependencies and information exchange among team members, in addition to summarizing and synthesizing different ideas during interactions. Salazar and Lant's study also showed teams with leaders possessing multidisciplinary breadth created more innovative outcomes than other leaders. Morton, Eigenbrode, and Martin's (2015) model of Architectures of Adaptive Integration, in turn, combined with insights from a primer on integration in large socio-ecological projects, highlighted the need for periodic adjustments.

Projects do not necessarily follow an ideal model of transdisciplinarity due to their heterogeneity. Their architecture varies as a result, though the authors identified six common attributes. They encompass goal-oriented structures, disciplinary and crossdisciplinary components, mechanisms for connecting, monitoring and adaptation processes, defined boundaries that are still flexible, and awareness of a project's Architectures of Adaptive Integration. Eigenbrode et al.'s (2017) primer further highlighted three structural challenges to integration: linking different conceptual frameworks to allow creative synthesis, using communication technology to enable relationship building and effective collaboration, and involving stakeholders in all stages of research to ensure knowledge is actionable. Explicit communication is essential for creating a shared culture that keeps team members focused, reduces uncertainty, and encourages and rewards collaboration. They also need to agree on rules of engagement, strategies for negotiating differences, and clear expectations for data management and handling publications and other products. Pohl et al.'s (2010) case study of sustainability projects further revealed the role of thought styles, a concept attributed to Ludwig Fleck (1979).

The projects focused on Kenya (vulnerability to drought), Switzerland (soil protection), and Bolivia and Nepal (conservation vs. development). Three challenges of integration appeared: addressing power relations, interrelating different perspectives, and promoting a negotiated orientation. Researchers are often skeptical about rigor while practitioners and stakeholders are skeptical about the practical relevance of academic knowledge. In the Swiss case, for example, farmers were interested in producing food, government agencies in protecting soils, and soil scientists in generating theories about soil functions and processes. Communication fostered growing trust and farmers' interest in science, including the possibility of cross-breeding improved seeds of a hybrid maize with the local variety. Researchers provided results, though were often handing back information collected from the agro-pastoral community's experiential knowledge about drought and mitigation.

Pohl and his co-authors also identified three roles researchers play in co-producing knowledge. The first—*Reflective Scientist*—provides scientific expertise, privileging academic perspective. The second—*Intermediary*—mediates between different thought styles, in this case moving across scientific, administrative, and farming collectives. The third—*Facilitator*—enhances communicative processes between thought collectives, promoting openness and joint reflection for common understanding and collective action by initiating a collective learning process. Roles differed, though, across the four projects. The Swiss case, for example, exhibited a high degree of researchers' implicit authority, while scientific expertise loomed larger in Nepal. As a result the researcher remained in the role of scientist. Comparison of projects also suggested the challenge of power is best addressed by researchers in the role of a *Reflective Scientist*, whereas challenges of integration and sustainability can be addressed by either researchers or other actors functioning as an *Intermediary* or *Facilitator* of a learning process. In the case of drought in Kenya, for instance, exploring and validating farmers' knowledge as a thought style helped explain why research and transfer of expert knowledge had little or no influence on agropastoral practices.

Public Engagement

Malin Mobjörk (2010) distinguished transdisciplinarity in general from participatory transdisciplinarity. The latter hinges on whether societal actors are in a secondary role as consultants responding and reacting to research or are active agents in co-producing knowledge. Pohl et al.'s previous example calls attention to public engagement, though Tebes and Thai (2018) cautioned understanding its nature is complicated by the variety of names associated with the concept. Major examples include participatory action research, empowerment evaluation, citizen science, patient and public involvement in research, stakeholder engagement in research, and community-engaged or community-based participatory research. Here, as elsewhere, contexts differ. Purposes also range from product invention in the industrial marketplace to democratic solutions to problems communities are facing to new protocols and treatment regimens for practitioners and patients. All of these examples aim to involve users and other stakeholders, a prime objective in transdisciplinary problem-oriented research. However, academic perspectives continue to dominate across sectors in all contexts of problem solving. Mercy Brown-Luthango (2013) cited, in particular, the Philippi CityLab in South Africa. Four thematic labs focused on urban flooding, the relationship between urban health and the built environment, urban ecology, and climate change. The flow of knowledge, however, was primarily outward from faculty and students. Researchers at the Institute for Socio-Ecological Research in Germany also base transdisciplinary research on translating a societal problem into an academic research question and epistemic object. This step is not limited to scientists. Jahn, Bergmann, and Keil (2012) emphasized it proceeds in cooperation with societal actors, and outcomes are integrated back into both societal and scientific contexts. However, other initiatives foster greater community engagement, facilitated by four boundary spanning roles Weerts and Sandmann (2010) identified: encompassing *Problem Solvers*, *Technical Experts*, *Internal Advocates*, and *Champions*.

Problem Solvers are close to a community and have a task orientation that is primarily technical and problem-based. *Technical Experts* are typically tenure-track faculty with a high degree of disciplinary expertise, so are content specialists. *Internal Advocates* also tend to be faculty in disciplines and occupy leadership positions focused on tasks of developing infrastructure. *Champions*, for their part, are integrated within a community and exhibit both socio-emotional and leadership orientations with external focus. These roles are not absolute, however. Weerts and Sandmann found individuals may move in and out of them prompted by changes in responsibilities, expertise, and position in partnerships or their universities. Community engagement received a boost in 2006 with the US National Institutes of Health's Clinical and Translational Science awards. Yet, a 2012 inventory revealed few community members were integrated into research teams. They were more often advisers or relegated to segmented activities. Fewer than 10% of awards involved them in ethics and clinical interactions. Moreover, evaluation does not typically include community metrics such as policy changes, improvements in health, and new partnerships. Power also differs between academic leaders and community representatives, the latter of whom often distrust academia and are insufficiently funded.

Patient engagement is a parallel example. It ranges in Sherry Arnstein's (1969) ladder from illusory participation or tokenism to meaningful influence in decision-making and bidirectional learning. A project to revitalize older residential neighborhoods on outskirts of Quebec City illustrates the top rung of Arnstein's ladder. Després, Fortin, Joerin, Vachon, Gatti, and Moretti (2008) built on Jürgen Habermas' (1987) theory of communicative action in a nearly year-long project combining scientific and action research with participatory design. Scientific and academic knowledge alone, they admonished, cannot deal adequately with the complexity of revitalizing neighborhoods. It is aimed at "what is true," but other forms of knowledge are also needed: instrumental ("what is possible"), ethical ("what is good"), and aesthetic ("what is beautiful"). Stakeholders confronted the four types of knowledge in encounters allowing representatives to express their views. In the process a fifth type emerged, a hybrid generated by intersubjectivity of "making sense together" in a form of "communicative rationality" (Després et al., 2008, p. 21).

Engagement of stakeholders also has implications for the meaning of translation. The state-of-the-art report on team science aligned the concept with application and transfer (NASEM, 2015). This narrow equation, however, ignores a large body of scholarship in humanities and translation studies on material, textual, and cultural factors that belie a universal original meaning. Moreover, Dvora Yanow (2004) noted, translators are not passive conduits. Echoing Akkerman and Bakker (2011), boundaries are dialogical phenomena. Hence, they require negotiation of meaning. In social psychology, the concept of the human mind as a unified subject is challenged by realization that individuals have multiple and even discontinuous identities. Transfer, Akkerman and Bakker added, is mostly about one-time and one-sided transitions, while boundary crossing is ongoing with two-sided interactions invoking Mikhail Bakhtin's (1981) concept of a heteroglossia of languages. It reinforces the difference between literature on transfer, Akkerman and Bakker emphasized, as well as literature on boundary crossing and boundary objects. The latter focuses on ongoing interactions and multiple contexts, while valuing diversity as a resource rather than something to minimize or avoid. Paul Carlile (2002) further observed pragmatic trade-offs are required.

Simply transferring knowledge across a syntactic or information-processing boundary and translating meanings at a semantic or interpretive boundary do not suffice. Neuhauser, Richardson, Mckenzie, and Minkler (2007) identified phases in evolution of translational approaches in health research that are not restricted to that domain. The first models, extending from 1960 to the mid-1990s, were linear. They depicted knowledge as a product transferred from researchers to users, characterized popularly as bench-to-bedside, with direct dissemination to clinical and other applications. In contrast, the second phase is characterized by exchange, recognizing knowledge is produced and integrated through relationships of collaborators and stakeholders. The third and current phase emphasizes integration in a whole-system approach factoring in priorities, cultures, and contexts. It also parallels transdisciplinary action research models and integration and implementation science.

Even so, step-phase models remain prominent. Hall, Vogel, Stipelman, Stokols, Morgan, and Gehlert's (2012) model of team science identified four stages in which stakeholders may be involved, beginning with *Development* and *Conceptualization*. Participation in project framing during *Development* and *Conceptualization* are important, though active integration is often emphasized in latter stages. At the third phase of *Implementation*, to illustrate, Latina community health educators might help identify study sites, recruit participants, or develop intervention strategies and materials. Or they might help collect and analyze data about clinical services. During the fourth stage of *Translation* stakeholders might share experiences from a community trial in a conference, or groups of patients, caregivers, and investigators might help develop guidelines for chronic conditions. Yet, influence still typically flows from science outward. Increasing public engagement prompted Tebes and Thai (2018) to coin the term *participatory team science*. However, they cautioned, engagement at all stages is typically unrealistic, cost prohibitive, and unsustainable. Russell, Wickson, and Carew (2008) warned against idealizing the *agora*. It is a heterogeneous and complex space with its own imbalances of power.

Impediments to transfer and translation appear in many fields, though the Clinical Research Roundtable of the US-based Institute of Medicine recognized two major blocks in health research and patient care in particular:

- The first—T1—impedes transfer of scientific discoveries in the laboratory to new methods for diagnosis, therapy, and testing in human studies.
- The second—T2—impedes translation of clinical studies into clinical practices.

Commenting on the blocks, Steven H. Woolf (2008) noted their knowledge bases differ as well. T1, for example, requires mastery of molecular biology, genetics, and other basic sciences in order to grapple with biological and technological concerns, recruitment for clinical trials, and regulatory matters. In contrast, the laboratory for T2 is the community and care settings. This location, Woolf explained, introduces complex dimensions of human behavior and organizational inertia, as well as infrastructure and resources along with conditions requiring implementation methods in real-world contexts. Interventions also require knowledge of other fields: including epidemiology, communication theory, behavioral science, public policy, financing, organizational theory, system redesign, informatics, mixed-methods research, psychology and cognition, social marketing, economics, and political science. A T3 form of practice-based research, Woolf added, has also been proposed, as well as a T4 approach to global health. Yet, he admonished, even the difference between T1 and T2 echoes the complexity of working across sector boundaries, raising the question of how translation boundary work operates.

Translation Boundary Work

Fam, Smith, and Cordell's (2017) findings from interviews with transdisciplinary leaders captured key qualities associated with communication and collaboration across disciplinary and sector boundaries. Despite a small sample, the 14 informants were involved in areas as diverse as natural resource management and tertiary education. They also came from a wide range of locations in Australia, the US, Europe, and South America. Findings revealed six core qualities conducive to collaboration: curiosity, creativity, commitment, critical awareness, communication, and connectedness. Figure 4.1 highlights key aspects: articulating one's approach in an accessible manner, listening to others, negotiating differences, building consensus, and translating knowledge from research to end-users.

Fig. 4.1. Communication.
Source: Fam, Smith, & Cordell (2017).

Two additional examples underscore the centrality of communication and collaboration. Quick and Feldman's (2014) study of urban planning illustrated translation boundary work in a mid-size city. Urban planning relies on specialized language learned in academic settings. It governs zoning ordinances, action plans, funding, and performance measures. Stakeholders agreed to introduce nonresidential uses or to increase housing units in some areas. The city planner acknowledged guidelines on density needed updating but knew revising them would be complex and controversial. When residents asked her to use her expertise in zoning law and implementation to draft new code language, she suggested starting with a simple, standardized definition of *density* for the entire city. At an open community meeting, though, residents discussed how they would like their neighborhoods to feel and suggested instead a customized combination of density-control measures. The planner was willing to try because the community supported her.

Quick and Feldman described the code ultimately adopted as novel and demanding: it included eclectic combinations of density-regulation tools in each neighborhood that would require more sophistication to implement than a standard uniform code. Reflecting on the case, they identified three generic practices that enabled treating boundaries as junctures rather than barriers:

- Translating across Differences: Adopting another language or way of expressing understandings to create a new domain in a multidirectional fashion, eroding the boundary between technical and non(*sic*) expertise;

- Aligning among Differences: Recognizing the presence of distinctions and finding ways to enhance connections for pursuing new shared interests and stakes;
- Decentering Differences: Finding ways to work that displace the primacy of one entity over others, relocating the focus to areas for coordinating joint movement forward.

Bidirectional translation resulted in a more versatile and nimble code. Beyond immediate outcomes, Quick and Feldman added, the project promoted resilience, building capacity for coping with novel problems in the future, disruptions, and resource constraints. A stance of inquiry toward boundaries also facilitated being cognizant of barriers, questioning whether differences are important or fixed, and discerning possibilities for change.

The second and extended example is Fam and Sofoulis' (2017) case study of a waste management project in Alaska. The research team brought pertinent experience. One was a microbiologist with Indigenous and public health expertise. Consultants were also experienced in STEM, participatory design, and transdisciplinary projects, including arctic engineering of water and sanitation systems. Yet, no member of the team had formal qualifications in social sciences and humanities. Social research is often deemed crucial to solving complex problems. However, Fam and Sofoulis admonished plugging social data into a predictive model with technical, environmental, and economic information privileges positivist science. In contrast, interpretive researchers aim to get inside a situation by using bottom-up qualitative methods of fieldwork and observational studies, surveys, interviews, diaries, and photo journals. The project team spent two weeks talking with 25 participants from local villages, regional consortiums, and state-wide organizations. However, the call for proposals did not allow for in-depth community participation and engagement, even though villagers possessed pertinent expertise about their water and sanitation history, economic viability of alternatives, and long-term acceptability.

Even with shortfalls, the project exhibited innovative features. Qualitative boundary conditions and cultural parameters for design yielded insight into economic constraints, capacity to manage and operate a system, and preferences for water sources. Engineers and scientists also appreciated other kinds of knowledge. Yet, the budget prioritized conventions of engineering, reflecting, in Fam and Sofoulis' words, an epistemological monoculture. Design thinking that could have tailored decentralized solutions was countermanded by top-down Big Water logic in a one-size-fits-all approach to 300+ remote villages. In the end, the team opted for a workshop on a proposed design and faxed a questionnaire to community leaders for feedback on technical solutions. Projects in many other areas also end up delivering a final report, then the team disbands.

Despite falling short, though, Fam and Sofoulis identified three major translation devices observed in other projects as well:

- Visualization: A diagram functioned as a boundary object that helped visualize the relationship between technical and social dimensions as hardware (technologies of taps, toilets, treatment and recovery procedures) and software (social acceptance, preferences, institutional arrangements, local capacities, and skills for using, operating, and managing a system).
- Design Principles: A design consultant described the process as weaving between knowledges aided by principles that helped align needs, preferences, economic viability, and acceptance of community members. Social research data also yielded insights into cultural practices of water use and managing, operating, and maintaining local sanitation systems. This approach differed from the technical building-block model of engineering.
- Linguistic Shifts: Language played a significant role in bridging and translating across differences. The terms *hardware* and *software* helped convey the idea of an interwoven socio-technical reality, while the engineering concept of management prototyping conveyed a need for new technical systems to consider socio/technical innovations.

Together the three translation devices Fam and Sofoulis identified helped mediate and negotiate objects, generating a shared conceptual framework or plan of action. Here, as in other cases, language also shaped both cognition and behavior, influencing how collaborators think and communicate, while linguistic shifts signaled formation of interlanguage. Visualizing, as well, is a form of language based on imaging. Together the three translation devices suggest crossdisciplinary and cross-sector work are forms of creative design. Darbellay, Moody, and Lubart (2017) acknowledged interdisciplinarity, creativity, and design are objects of study in an autonomous field. Yet, they are intertwined. Growing awareness of design features also parallels a shift in management literature from focus on organizing teams and facilitating interactions to creating research cultures and behavioral dynamics. As we shall see in the next chapter, though, research partners and students do not necessarily possess the skills and competencies needed for communication and collaboration across boundaries.

5
Learning

> Understanding how people learn to synthesize is essential, if we are to design quality instruction and support learners to fulfill these institutional aspirations.
>
> Because syntheses vary, we must investigate the epistemological foundations on which learning to synthesize stands and attend to the common *and* idiosyncratic features of interdisciplinary syntheses and the concomitant criteria by which we might deem them acceptable.
>
> **Veronica Boix-Mansilla (2017)**

Learning occurs on multiple levels: from individuals, teams, and projects to fields and large, cross-institutional initiatives. Settings also vary. Paul Gibbs' (2015) collection of essays on transdisciplinary learning and practice, for instance, spanned curriculum, healthcare, applied sciences, and cross-sector partnerships. Moreover, not all formats are the same. The report on team science from the National Academies of Science, Engineering, and Medicine distinguished education in traditional classroom settings from training or professional development outside academic institutions (NASEM, 2015). The distinction is not air-tight, however. Heightened interest in teamwork has resulted in adding training modules, short courses, workshops, and summer residencies to the curriculum. Heeding the opening epigraph, this chapter investigates the nature of learning across boundaries. "All learning," Akkerman and Bakker observed, "involves boundaries" (2011, p. 132). However, they take on greater magnitude in crossdisciplinary and cross-sector work.

The chapter begins with insights on integrative learning in interdisciplinary education, which has a longer history than transdisciplinary education. It is grounded in a shift from content-based to process-based integration, the theory of constructivism, and the concept of reflective equilibrium. After noting parallels with transdisciplinary projects and education, it examines the nature of social learning, framed by four theoretical discourses for interdisciplinarity and interprofessionalism bridging occupational

Beyond Interdisciplinarity. Julie Thompson Klein, Oxford University Press. © Oxford University Press 2021.
DOI: 10.1093/oso/9780197571149.003.0006

groups: *Communities of Practice*, *Critical-Historical Activity*, *Complexity Science*, and *Actor-Network Theory*. Turning specifically to transdisciplinary work, the chapter then defines the nature of a communication boundary space and related concepts, in the process accounting for a shift from identification to transformation while incorporating complementary concepts of double- and triple-loop learning, reflexivity, convergence, transactivity, and heuristics. Finally, after identifying individual and collective competencies and expertise in both education and training, the chapter draws insights from two cross-sector case studies, beginning with a sustainability project in the East India Plateau that bridged academic and community knowledge, followed by healthcare in two hospital settings that bridged disciplinary and occupational expertise.

Interdisciplinary Learning

The place to begin is interdisciplinary education because it has a longer history of focus on learning than literature on transdisciplinarity. Fiore, Gabelica, Wiltshire, and Stokols (2019) tracked discussions of interdisciplinary learning back to Brozek and Keys' (1944) description of interdepartmental research programs such as the Yale Institute of Human Relations in the 1930s. Related topics at the time included communication and group process, conflict, and role ambiguity. The authors also highlighted development of interdisciplinary programs in health sciences during the 1960s but did not cite earlier warrants for integrative learning and, subsequently, experimental programs during the late 1960s and early 1970s that were laboratories for integrative pedagogy and learning. A deeper search of pertinent literature reveals several historical shifts that foreshadowed current emphasis on process in both inter- and trans-disciplinary research. One of the major developments occurred in thinking about integration among educational groups.

Edward Ciccorico (1970) recounted a major change among participants in a 1935 meeting sponsored by the US-based National Education Association and a 1937 book on integration. Concluding that complete unity is impossible, they proposed instead the concept of unifying rather than unified approaches. Subsequently, members of a 1948 workshop sponsored by the Foundation for Integrative Education distinguished content integration, which bridges subjects, from process integration, which emerges in the interplay of an individual and an environment. They also differentiated integration as a synthesis of accepted postulates from integrative building of new conceptual modes. Boix-Mansilla and Lenoir (2010) linked integration and interdiscipinarity with other movements as well. The Herbartians' doctrine of correlation, for example, went beyond the doctrine of concentration on specific subjects alone. The Progressivist movement promoted development of the whole child; John Dewey's emphasis on a child's interests went beyond a solely disciplinary focus; and campaigns for an integrated core curriculum and the general education movement fostered integrative learning at all levels.

Combining literature review and hands-on work with teachers, Boix-Mansilla (2017) proposed pragmatic constructionism is the heart of interdisciplinary learning. It is purposeful and provisional, as well as a pluralist embodiment of multiple forms (e.g., aesthetic, analytical, interpersonal, and ethical understanding) and multiple symbol systems (e.g., mathematical, visual, linguistic, and kinesthetic). She further linked interdisciplinary learning with Catherine Elgin's (1996) concept of reflective equilibrium. Separate components are weighed in light of each other and prior assumptions, thereby scaffolding understanding. To reiterate, more attention has been paid to individual than to group learning, but increased attention to collaboration is casting greater light on the social nature of learning.

When Brown, Collins, and Duguid (1989) identified salient features of group learning, they listed collective problem solving, displaying multiple confronting ineffective strategies and misconceptions, as well as providing collaborative work skills. All forms of group learning are not the same, however. The terms *cooperative learning* and *collaborative learning* are sometimes used synonymously, but Michael J. Baker (2015) made a crucial distinction. Cooperative learning typically entails divisions of labor toward a common goal. In contrast, collaborative learning involves joint attention, communication, interaction, mutual engagement, and co-elaboration of knowledge. In identifying common traits of interdisciplinarity and interprofessionalism Angus McMurtry (2013) reported that educators are increasingly using the terms *collective* and *socio-material* in describing theories of learning. Their boundary work not only engages multiple disciplinary specialties. It also fosters a holistic approach to healthcare by teams spanning occupational professionals. McMurtry did not mention transdisciplinarity. However, he framed learning around four theoretical discourses cited in literatures on both inter- and trans-disciplinarity.

The first theoretical discourse is *Communities of Practice* (CoPs). To recall Chapter 1, Lave and Wenger (1991) proposed CoPs to account for groups with a shared interest. As individuals participate in a community, knowing and identity become located in social relationships. Moreover, traits associated with interdisciplinarity appear in interprofessional practice. Diverse perspectives come into play when constructing common ground and negotiating conflict across specialties, with the aim of synthesis. The second discourse—*Cultural-Historical Activity*—also situates learning in social processes and collectives. McMurtry traced the theory to Vygotsky, Ilyenkoc, and other Marxist-influenced thinkers who argued for activity systems as the primary locus, factoring in a mix of participants, tools, rules, divisions of labor, and common objects.

Here, as well as in the third and fourth discourses, perspectives require negotiating conflict. A learner-centered view of the third discourse—*Complexity Science*—has become a cornerstone of theorizing both inter- and trans-disciplinarity. The 2015 NASEM report on team science, for example, viewed challenges of communication and coordination in collaborative research through the lens of complexity. In a nutshell, the aggregate behavior of a system emerges from intersections of its parts in a network of non-linear interactions, producing a form of self-organization and emergence. The fourth discourse—*Actor Network Theory*—locates learning in relationships and interactions that emerge within activity networks composed of nodes and ties where interactions and connections occur. Parallel tenets of process-based integration, constructivism, and social learning appear in literature on transdisciplinarity as well, and here again conflict is a factor. Decuyper, Dochy, and Van den Bosschec (2010), in fact, called it a major barrier to team learning in transdisciplinary contexts, though constructive conflict can create dialogical space where discordant processes coexist. Models of integrative process depict how differences are bridged in addressing a shared challenge.

From Identification to Transformation

Based on 17 years of transdisciplinary praxis at the University of Technology Sydney's Institute for Sustainable Futures, Mitchell, Cordell, and Fam (2015) defined three transdisciplinary outcome spaces that create focal points for dialogue. The first space—*Situation* or Field of Inquiry—is a site for improvement in everyday contexts of collaborators and stakeholders. The second—*Knowledge*—comprises stocks and flows in academic and other forms of knowing that shape work in both scholarly and societal contexts. The third—*Learning*—beckons mutual or transformative capacity, though definitions vary. Decuyper, Dochy, and Van den Bosschec (2010) identified 30 connotations of team learning in organizations, with differing emphasis on processes, structures, outcomes, environments, and subsystems. Even so, mutual learning is widely regarded as essential for collaborative transdisciplinary research. Golja, McClean, and Jordan (2018) even called it the basic principle of transdisciplinarity.

Knowledge is not simply exchanged but is co-constructed as individuals with differing views and stakeholders work together. Thus, it is a process of social learning Schauppenlehner-Kloyber and Penker (2015) associated with interactions that foster co-creation of knowledge. At a finer-grained level of detail Akkerman and Bakker (2011) identified four mechanisms for learning that parallel key traits of integrative process:

- *Identification* is a dialogical process that reveals differences when attempting to construct and reconstruct boundaries.
- *Reflection* entails meaning-oriented learning processes that render participants' perspectives, practices, and identities visible.
- *Coordination* enables cooperation, including use of boundary objects to facilitate communicative connection through translation.
- *Transformation* leads to more significant changes, including the possibility of a new boundary practice.

Identification is a widely recognized first step for collaboration, entailing clarification of individual perspectives for comparative discussion. *Reflection* deepens learning of relevant knowledge and reflection on assumptions about self and others, thereby enlarging and even changing perspectives. Bergmann, Jahn, Knobloch, Krohn, Pohl, and Schramm (2010, 2012) described reflexivity as both transgressive, challenging dominant assumptions and power structures, and transcendent, engaging epistemological flexibility and synergies between contingent disciplinary and cultural knowledge. *Coordination*, in turn, fosters communicative connections that can increase boundary permeability and routinize new approaches, while *Transformation* connotes more profound changes, such as hybridizing elements from different contexts. A new in-between practice or interdisciplinary field might even emerge, but that is not a routine outcome. More often individuals factor new knowledge into their practices without sacrificing original expertise. Movement from *Identification* to *Transformation* also invites parallels to double- and triple-loop learning.

Argyris and Schön (1996) associated single-loop learning with assessing performance against established rules or standards. In contrast, they are questioned in double-loop learning and may even result in major changes to strategic choices and objectives. Argyris and Schön did not mention triple-loop learning explicitly. However, it connotes a higher or deeper level of learning that occurs when processes or methodologies generate new principles or paradigms (after Tosey, Visser, and Saunders, 2011). Hence, it is also associated with transformational team learning, calling to mind Jack Mezirow's (2003) definition of transformative learning as a discourse that moves beyond fixed assumptions. By rendering them more inclusive, reflective, and capable of change, it also enacts critique. Double- and triple-loop learning underscores the centrality of reflection as well. Many associate them with Donald Schön's (1984) concept of reflection-in-action. They highlight a reflexive epistemology of practice that develops in the course of thinking. Thus they are pragmatic, situated, and elevate tacit knowledge.

In a study of individual and distributed cognition in interdisciplinary teams, Derry, DuRussel, and O'Donnell (1997) integrated the concept of situation awareness with understandings of mental models as well. In a separate study of an interdisciplinary team focused on STEM education, DuRussell and Derry (2005) echoed their findings. Prior knowledge helps shape individuals' mental models. As situational learning occurs, though, new models influence the content and the structure of their schemas. Comparably, Hauke Riesch (2010) observed people confronted with an unfamiliar concept typically try to anchor it in familiar representations. However, the original becomes modified as a new concept emerges. Harmony Bench (2018) illustrated the constructive pragmatism of situated learning in a classroom exercise for teaching ballet that has generic import. *Enchaînement* involves arranging codified movements and steps into a dance. Performance data delimit parameters. Yet, the creative practice entails filtering, sorting, fitting together, and displaying an object through experimental combinations. Connections are not pre-determined. Danielle Boutet's (1993) notion of artistry, Jill Vickers (1997) suggested, provides a model of how to teach interdisciplinarity in an open field.

To recall Chapter 3, an open field does not necessarily settle on a canon, foundation, paradigm, or epistemology. Boutet conceived of interdisciplinarity as a reflective process that begins with knowledgeable borrowing. An artist, for example, unbinds tools, techniques, methods, generative theories, and materials from disciplinary packages. The working context is supplied by goals and frameworks the artist creates in order to mediate interaction of components. Extending unbinding and mediating to learning in an open field, Vickers argued students should be evaluated on the ground of generative process and their ability to legitimate a new conceptual framework, not a priori discipline-based criteria. Generativity also includes the possibility of grappling with the unexpected. Darbellay, Moody, Sedooka, and Steffen (2014) acknowledged the value of Allen Repko's 10-stage model of interdisciplinary research process in a textbook for students. Repko (2008) himself even mentioned the possibility of feedback, iterativity, and negotiation. Yet, they cautioned, his model follows a standard progression, from formulating a problem, questions, and hypotheses to making methodological choices and analyzing and interpreting results. It does not account for the unexpected.

Competencies

Once again, interdisciplinarity provides a starting point, though within a broad range of curricular formats from discipline-based courses to integrated proseminars. Even with differences in format, however, litanies of skills for both inter- and trans-disciplinary work bear a strong resemblance. Learning outcomes typically include tolerating ambiguity, demythologizing experts, seeing new and different questions and issues, and drawing on multiple methods and knowledge. Table 5.1 is Spelt, Bieman, Tobi, Luning, and Mulder's (2009) roster of subskills and conditions that span disciplinary knowledge, understanding of interdisciplinarity, its balance with disciplinarity, personal characteristics conducive to integration, and explicit attention to learning process. They focused primarily on individuals, though added the pedagogical goal of fostering collaboration and teamwork in curriculum development and teaching.

Although inclusion of disciplines reflects a widely held belief they comprise the foundation for interdisciplinary work, when identifying learning outcomes for graduate-level science and engineering education Borrego and Newswander (2010) went beyond knowledge and basic competence in a discipline. They added integration, breadth, critical awareness, and communication as well as capacity for teamwork. Many studies continue to separate individual and group learning. Yet, Decuyper, Dochy, and Van den Bosschec (2010) maintained their relationship is dialectical. Comparably Volet, Vaurus, and Salonen (2009) cautioned against reducing analysis to either level. Both are required. Rebecca Freeth emphasized the importance of personal capacity for teamwork in transdisciplinary education (pers. comm., 12 November 2018). That said, literature on team science accentuates skills needed for collaboration. Fiore, Gabelica, Wiltshire, and Stokols (2019) joined others in starting with knowledge of disciplinary content but added capacity to engage in teamwork. Related interpersonal competencies include active listening to others, asking them to explain what is meant if not transparent, communicating one's own meaning, and building rapport with others.

Swasan Khuri's (2018) litany of core competencies for team science emerged from her review of literatures on inter- and trans-disciplinary scholarship, non-profits, and businesses. Khuri identified seven core competencies

Table 5.1 Overview of Potential Subskills and Conditions

Potential subskills and conditions for interdisciplinary higher education		
Interdisciplinary Thinking	Having knowledge	Knowledge of disciplines
		Knowledge of disciplinary paradigms
		Knowledge of interdisciplinarity
	Having skills	Higher-order cognitive skills
		Communication skills
Student	Personal characteristics	Curiosity
		Respect
		Openness
		Patience
		Diligence
		Self-regulation
	Prior experiences	Social
		Educational
Learning Environment	Curriculum	Balance between disciplinarity and interdisciplinarity
		Disciplinary knowledge inside or outside courses on interdisciplinarity
	Teacher	Intellectual community focused on interdisciplinarity Expertise of teachers on interdisciplinarity Consensus on interdisciplinarity
		Team development
		Team teaching
	Pedagogy	Aimed at achieving interdisciplinarity
		Aimed at achieving active learning
		Aimed at achieving collaboration
	Assessment	Of students' intellectual maturation
		Of interdisciplinarity
Learning process	Pattern	Phased with gradual advancement
		Linear
		Iterative
		Milestones with encountering questions
	Learning activities	Aimed at achieving interdisciplinarity
		Aimed at achieving reflection

Source: Spelt, Bieman, Tobi, Luning, and Mulder (2009)

and associated behavioral patterns. Echoing other rosters, it includes effective listening and adaptation to others. However, she also called attention to the importance of monitoring and presenting oneself as well as practicing communication ethics, including expressing messages clearly for others to understand, and more meta-level identification and explanation of communication processes fundamental to successful collaboration. In defining characteristics of trans-disciplinary individuals, Tanya Augsburg (2014) bridged individual and group capacities. She affirmed Bruce, Lyall, Tait, and Williams' (2004) roster of qualities associated with individuals, such as curiosity, flexibility and adaptability, and the ability to absorb information. Augsburg added, though, reflection on integration, risk-taking, and three requirements for successful collaboration: willingness to learn from other disciplines, good communication and listening skills, and an ability to work with others. Like others, Augsburg also included a creative mind.

In defining the relationship of creativity and interdisciplinarity, David Sill (1996) highlighted three aspects in Getzels and Csikzentmihalyi's (1964) classic definition: original production, cognitive problem solving, and subjective experience. He also cited relevant concepts in Arthur Koestler's (1964) groundbreaking work. Bisociative thinking derives from synthesizing independent matrices of thought, while creative tension requires resolution when two or more of them conflict. More recently, alignment of crossdisciplinarity with innovation has heightened interest in creativity. A survey by the American Management Association identified it as one of the 4Cs of 21st-century skills students need, along with communication, collaboration, and critical thinking (AMA Staff, 2019). A 2013 report entitled *Arise 2* also called transdisciplinary opportunities a common theme of research in science and engineering. They claimed interdisciplinarity only borrows techniques from different fields without integrating them, ignoring a sizable literature on integration in interdisciplinary research and education. Yet, *Arise 2* captures the power of deep integration in transdisciplinarity for both basic discovery and application (American Academy of Arts and Sciences, 2013). Despite this momentum, however, individuals often lack a sense of belonging.

Felt, Ingelsböck, Schikowitz, and Völker (2013) adapted the concept of epistemic living spaces to highlight the need for guiding maps to both explicit rules of practice and tacit assumptions about practice and behavior. They illustrated the need in an Austrian doctoral program entitled Transdisciplinarity as Culture and Practice. Some students were uncertain about how to navigate complex dimensions of challenges such as sustainability. As a result, they defaulted to a familiar disciplinary map with lower risk and uncertainty. Reporting on another doctoral program at Stellenbosch University in South Africa, Van Breda, Musango, and Brent (2016) proposed transdisciplinary epistemic communities are network-like structures with dyadic relationships and nodes between academic and social actors. They based this premise on three projects: food security for subsistence farmers at the micro-household level in rural areas of the Kwa-Zulu Natal region, sustainability of renewable energy technologies improving technology assessment practices in the Eastern Cape Province, and experiences reconnecting humans and nature. This case affirms the importance of translation. In the case of food security, the researcher had to treat farmers' tacit or experiential knowledge of equal value to theoretical knowledge.

An example from a challenge-driven tertiary institution in Australia also illustrates the importance of creating a safe space for students to learn from failure. Fam, Lopes, and Mitchell (2021) recounted a Living Laboratory experiment to pilot an alternative sanitation system in a multi-story campus building. Despite the university's commitment to real-world problems and cross-sector partnerships, the project fell short of creating a platform for large-scale trials. Academic resistance to participatory decision-making impeded progress as well as competing interests of researchers who returned to their regular homes afterward. Furthermore, occupational practitioners, such as the plumber, had their own professional regulations and codes to follow. The experiment was not a complete failure, however. It launched new research, while modeling a place, in the authors' words, to fall over and to skin your knees without suffering penalties. BinBin Pearce's (2021) account of learning-to-fail-forward described a safe place in a required course of the Department of Environmental Systems Science at the Swiss Federal Institute of Technology in Zurich. By cultivating curiosity and inquiry over certainty and correctness, it incorporates productive failure, treating problems as opportunities for learning and improvement.

Cultivating Competencies

Here again, discussion starts with interdisciplinary education. Deborah DeZure's survey of instructional methods revealed multiple "productive pedagogies" (2017, p. 558). Her roster included team-based projects and case studies; role-playing, simulations, and gaming; problem-, discovery-, and inquiry-based learning; field experiences such as internships and service learning; relational and feminist pedagogies that foster inclusivity; portfolios that collect and reflect on experiences; and high-impact practices such as first-year and capstone courses, learning communities, writing-intensive courses, and undergraduate research programs. Productive pedagogies also demonstrate Davidson and Janz's (2019) contention that specialists or experts are no longer the only active producers of knowledge. Boundaries between teachers and students give way to collaborative possibilities. The classroom is not the only site of productive pedagogies, either. Fiore, Gabelica, Wiltshire, and Stokols (2019) reported similar approaches in team science training that move beyond traditional forms of education. The majority of interdisciplinary experiential learning, they reported, comes through a range of coursework, internships, summer and in-semester research experiences, and mentoring. Strategies also include laboratory research and journal club meetings and products.

Fiore, Gabelica, Wiltshire, and Stokols further cited a number of grant programs. One in particular was targeted at the post-graduate level when students are entering the scientific workforce. The National Cancer Institute's Transdisciplinary Research on Energetics and Cancer I (TREC I) initiative aimed at developing competencies enabling teamwork. TREC centers supported a variety of training activities, including trans-disciplinary research courses, journal clubs, and retreats to develop skills in collaborative writing and research. They also included co- and multi-mentoring to expose trainees to multiple disciplinary perspectives. Federal programs have focused more specifically on developing sophisticated mentoring strategies to foster development of interdisciplinary career paths. Tallying strategies and models, Fiore, Gabelica, Wiltshire, and Stokols identified a set of core features of training for teams that are transportable across contexts of teams and tasks (Table 5.2).

In an earlier overview of trans-disciplinary training for team science, Justin Nash (2008) also reported knowledge and skills are best developed by combining formal didactics, mentoring, and research experiences. Figure 5.1 depicts multiple factors that influence outcomes, including both antecedent factors and intervening processes. Nash did not stop at a single point in time, when antecedents are set in motion. Formal mechanisms such as

Table 5.2 Core Features Needed to Examine the Effectiveness of Learning Approaches for Competencies Training

- Use small groups.
- Provide immediate and regular feedback (from instructors, peers, or professional guests).
- Use formative assessment allowing for refinements and adjustments of ongoing work.
- Ensure regular communication of ongoing work.
- Use complex and real-world problem solving.
- Support metadiscussion (via metacognitive techniques).
- Ensure multi-interactions among learners and instructors.
- Provide explicit training of problem solving skills.
- Use scaffolding/modeling from the trainer.

Source: Fiore, Gabelica, Wiltshire, and Stokols (2019)

presentations, publications, grants, and mentoring support outcomes over a five-year period and beyond.

Recalling their own experiences in both crossdisciplinary and cross-sector programs projects, Fam, Smith, and Cordell (2017) reported participants articulated the need for more than pragmatic skills. Attitudes and dispositions also shape mindsets. Misra, Stokols, and Cheng's (2015) concept of transdisciplinary orientation acknowledges their role in a composite picture of values, attitudes, beliefs, conceptual skills and knowledge, and behavioral repertoires that predispose an individual to collaborating effectively in teams.

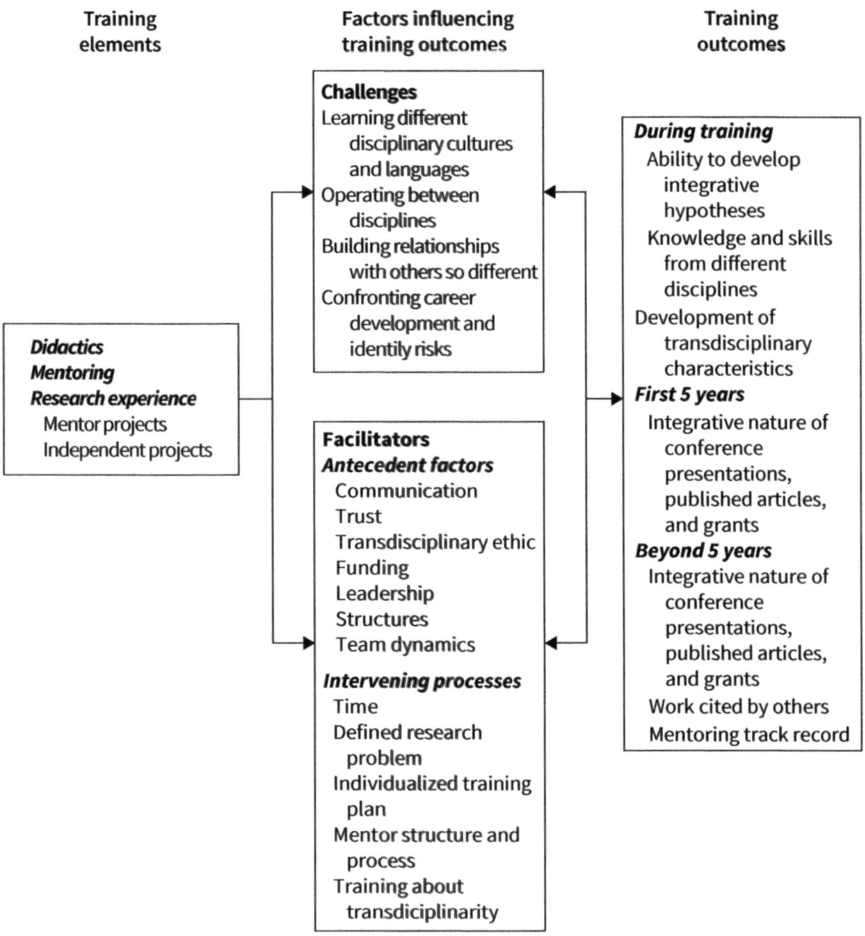

Fig. 5.1 Transdisciplinary Training Elements, Facilitating and Challenging Factors, and Outcomes.
Source: Nash (2008).

It encompasses both intrapersonal and interpersonal as well as intellectual dispositions individuals bring to a group.

TD Values are guiding principles that incline individuals to work with others while incorporating theories and methodologies from other fields into one's own research.

TD Attitudes are conducive to crossdisciplinary scholarship, including willingness to learn about other fields and to adjust disciplinary schema to fit demands of teamwork.

TD Beliefs pertain to benefits and costs of teamwork as well as views of theoretical and empirical discoveries and translational outcomes.

TD Conceptual Skills and Knowledge include an ability to view problems and questions holistically and to integrate concepts or methods from different disciplines.

TD Behaviors include habits such as reading journals and books from different fields, attending meetings outside one's field, and engaging with colleagues from other disciplines.

Reflecting on this set of dispositions Nurius and Kemp (2019) grouped values, attitudes, and beliefs into a cluster that may emanate from both personal characteristics and socialization in a discipline. Personal habits of mind include openness and respect for differences, including diversity of others' backgrounds. Intellectual curiosity and willingness to assume risk are also forms of openness that foster creativity, enhanced by both reflexivity and systems thinking. Self-assurance matters as well, without being defensive, while commitment to a project and its team members reflects a willingness to address challenges of crossdisciplinary and collaborative work, including tolerance for uncertainty. Interpersonal competencies, in turn, include an ability to engage in communication, relationship building, and collaboration. And knowledge-based competencies encompass both disciplinary depth and an ability to identify other disciplinary content relevant to a particular problem, with the aim of developing a shared vision or model. Successful collaboration also depends on a transactive memory system. Decuyper, Dochy, and Van den Bosschec (2010) defined transactivity as the degree to which learners acquire shared knowledge in the form of mental models or cognition and knowledge of who-knows-what. Team learning thus fosters a group-level system for encoding, storing, and retrieving information.

In a recent genealogy of the field of social ecology, Daniel Stokols (2017) further aligned transdisciplinary orientation with four additional Ts that students need today. Research in the 21st century, he exhorted, is increasingly transdisciplinary, transnational, team-oriented, and transcultural. Here, as in the prior chapter, leadership and management also require Walker and Nocon's (2007) notion of boundary-crossing competence for managing and integrating discourses and practices. This goal lies at the heart of the INTEREACH subgroup of the International Network for the Science of Team Science. INTEREACH aims to develop a new profession that coalesces traits of an Interdisciplinary Executive Scientist, specialists in integration and implementation sciences, and a variety of research development officers (https://www.intereach.org).

Related expertise, Hendren and Ku (2019) explained, lies at the interstices of cross-boundary communication and boundary negotiation, project leadership and management, and research and outreach to constituents who include both academics and external stakeholders. Tools and strategies for translating across forms of expertise and knowledge of integrative processes and synthesis are also needed. The position of Interdisciplinary Executive Scientist in particular includes epistemic fluency in the type of scientific problem a team is addressing along with skills of integration and collaboration. Individuals serving in this role, however, report having responsibility but not authority and recognition in the academic reward system, compounded by diminished time to advance other career needs, such as publishing in their personal specialties. Moreover, positions are often ad hoc, so not sustainable on a long timescale.

The challenge of going beyond short-term outcomes haunts training programs as well. Vogel et al. (2012), though, reported the National Cancer Institute's initiative TREC I yielded evidence of potential for long-term reorientations, based on analysis of training for TREC centers. Illustrating the importance of judging success by more than traditional proxy indicators such as publications, presentations, and grants, the indicators included changes in attitudes toward working across disciplines (in the form of an enhanced trans-disciplinary orientation), improved ability to work across disciplines (marked by gains in intrapersonal/interpersonal competencies for collaboration), and increased competency to work with other disciplines (achieved by development of both personal and scientific skills for trans-disciplinary research). In addition Vogel and colleagues reported increases in scholarly productivity for trainees who were affiliated with TREC. Multi-mentoring experiences were associated with greater transdisciplinary orientation overall, affirming the importance of learning opportunities beyond formal classroom didactics. This chapter, like the previous one, ends by extending understanding to cross-sector work in situ.

Learning in Situ

The kind of learning being sought is not a particular school or theory, Mitchell, Cordell, and Fam (2015) emphasized. It evolves in the course of working together for a common purpose. Examples abound. To recall the project on soil protection in Switzerland from Chapter 4, self-reflexive learning occurred as researchers' understanding of farmers' local knowledge increased. Villagers also had an opportunity at meetings to contribute ideas, then take home alternatives for collective reflection on vulnerability of their livelihoods and avenues for improving its security (Pohl et al., 2010). In a parallel project at a small, resource-scarce teaching-intensive institution in Puerto Rico, Medina, Báez, and Mèndez (2018) emphasized the importance of continuous capacity building and collaborative readiness. They also linked capacity building with efforts to promote shifts in the power structure of different sectors. Janet McMillan (2009, 2011) emphasized the importance of universities in particular to developing requisite skills to work across sectors. Her immediate example was service learning in South Africa, but individuals who have juggled academic and commercial expectations will empathize with the challenge of brokering. McMillan herself built on the concept of situated learning, post-Vygotskian theory of learning as a social practice, and activity theory.

As other examples in this book also demonstrate, the interface of higher education and society is a site of negotiation. Situated learning and communities of practice, McMillan emphasized, come together when sustained engagement occurs. Drawing insights from Star and Griesemer (1989), Bowker and Star (1999), and Wenger (1998) as well as Tuomi-Gröhn and Engeström's (2003) book on classification, McMillan expanded the framework by incorporating boundary zones, boundary objects, and brokering. Tuomi-Gröhn and Engeström, in particular, acknowledged such spaces are generally places of challenge, contest, and power relations. However, they added, intersections can be sites for learning and producing new knowledge. They are also spaces where communities of practice can reflect on their own discourse, structure, norms, and roles. Thus elements from more than one system are co-present. A boundary, McMillan concluded, is a point of intersection located between two activity systems. When McMillan was joined by Goodman and Schmid in 2016, they drew further conclusions from a project sponsored by the Knowledge Co-op at the University of Cape Town. It tested tools for analyzing the boundary work of complex university-community interactions.

McMillan, Goodman, and Schmid invoked Michael Gibbons' (2005) contention that the metaphor of *transaction space* shifts from translation across boundaries to dialogue. Universities, he exhorted, must cultivate knowledge and skills required for engagement. Service learning in particular, they added, has been called a form of border pedagogy informed by critical pedagogy and postmodernism as well as activity theory. Echoing Henry Giroux's (1992) book *Border Crossings*, they too situated boundary crossing in the politics of education. From a historical perspective, the authors recalled, community engagement in higher education in South Africa was promoted by an education white paper that admonished the need for transformation to contribute to post-apartheid South Africa. The document, they added, echoed Ernest Boyer's influential call for broadening definition of scholarship beyond research to include scholarship of teaching, application, and integration in a form of applied academics. Greg Cronin (2014) argued transdisciplinarity should also be rewarded in keeping with Ernest Boyer's (1990) notion of scholarship as discovery, especially when grappling with wicked problems.

The Knowledge Co-op at the University of Cape Town aimed to bridge boundaries of scholarship, teaching, and community engagement based on the European science shop model, which provides community groups with skills and resources available in the University. Student work in the Co-op counts toward an academic qualification and provides staff with a framework for research and student training and learning grounded in project-based work. Past projects have included studies of women awaiting radiation after breast cancer surgery, the stigma and the behavior of sex workers around pregnancy and motherhood, alternative energy sources for pumping water in rural municipalities, and computer training in organizations. Solitary learning, McMillan, Goodman, and Schmid concluded, is not adequate for preparing all parties to grapple with complex and even wicked problems that require joint understanding, coordination, and alignment of perspectives in different activity systems and communities of practice. Mutual learning is required. Two additional projects depict learning across sectors. The first was a transdisciplinary agricultural project in Indonesia aimed at improving livelihoods for Indigenous farmers by intensifying and diversifying a cropping system. The second and lengthier case was located in two hospitals where boundary objects rendered disciplinary backgrounds and occupational practices more accessible and open to collaboration.

In the first instance, Bill Bellotti (2017) drew insights about cross-sector engagement from a project in the rural East India Plateau. The staple food in targeted communities was a monocrop of transplanted rice that was ill suited to areas where population pressure has become problematic for smallholder farmers along with variable climate. A program called Professional Assistance for Development Action and several Australian-based research organizations combined research and development objectives in three stages:

Problem Framing and Team Building: to create a common research object and to build a collaborative team;
Creating Actionable Knowledge: to engage multiple stakeholders and learning as a planned outcome;
Integrating and Applying New Knowledge: to implement new practices while intensifying learning experiences.

During the first phase development professionals and scientists attended a three-day workshop. Farmers were not represented initially, however, so were not parties to the initial stage of problem framing. Yet, the concept of transdisciplinarity provided a retrospective lens on strategies aimed at combining scientific and social practices with differing epistemologies. Developing a year-round crop-planning tool, for example, facilitated learning about land and water resources, while visual representation depicted impacts on crop intensity and diversity using the technology of Google Maps. Introduction of a more reliable aerobic and directly seeded rice also had implications for the role of women. They were traditionally menial laborers responsible for planting rice. In the course of the project, though, they came to see themselves as farmers while teaching other villagers as well. Farmers' participation also fostered government officials' learning through feedback. Here, as in other projects, however, outcomes were limited. In the end, team members did not achieve a common level of understanding transdisciplinarity, boundaries were not addressed explicitly, and boundary objects did not provide a focal point for collaboration. Numerous other studies echo the importance of continuous capacity building.

The second case study yielded parallel lessons in a Western setting, based on Oborn and Dawson's (2010) analysis of multidisciplinary teams (MDTs) in two hospitals dubbed Metropolis and Suburb. The MDTs were mandated in the UK as part of a national cancer plan. The teams involved specialists from surgery, oncology, nursing, pathology, and radiology, plus a general practitioner at Suburb. Oborn and Dawson identified three methods in a "virtuous circle" of participation, learning, and identification (p. 844). The first—*Organizing Discussions*—facilitated learning how to engage in interdisciplinary talk while negotiating group relations. The second—*Acknowledging Other Perspectives*—helped to interrelate expertise by highlighting discrepancies and fostering understanding of others' viewpoints. The third—*Challenging*

Assumptions—entailed negotiating and broadening meaning while weighing contradictory evidence. Boundary objects also facilitated exchange. When members of teams met together to discuss cancer cases and patient management, X-rays, pathology specimens, and protocols functioned as boundary objects that promoted learning across specialities.

Identity construction and identity work were also woven into the process. In a separate study, Leibowitz, Wisker, and Lamberti (2016) suggested academic identity development may itself be considered a learning trajectory with a social dimension that reflects both individual histories and mutual engagement with shared repertoires. In the case of the two hospitals, the more individuals participated in a virtuous cycle of participation the more they could work together. Individuals also adjusted their roles, despite traditional hierarchies of disciplinary and professional stereotypes of surgeons as leaders, pathologists as detailed analysts, and nurses as marginal caregivers. The general practitioner at Suburb, for example, negotiated identity as an innovative practitioner at the forefront of research. And, the pathologist at Metropolis facilitated learning from other specialties to help construct collective identity. In the latter role, by keeping track of past referrals and monitoring whether patients from pathology made it to the MDT list, she fostered systematic discussion of all patients, not just selected ones.

The pathologist's brokering role was apparent when two of three small breast lumps in a patient were biopsied. The pathology report suggested only one was malignant, but the radiologist believed all would be so considered testing all three unnecessary. One surgeon and a radiologist instead proposed a re-biopsy, using the more accurate technique of coring. A second surgeon, who respected the radiologist's clinical skills, also highlighted discrepancies between possible interpretations of results and treatment options. Dialogue then ensued on biopsy methods and features the radiologist highlighted on scans. For her part, the pathologist zeroed in on differences as opportunities for learning. Her understanding of others' judgments and priorities enabled her to position her knowledge in Oborn and Dawson's concept of a virtual new space where boundaries could be made explicit. The third method—*Challenging Assumptions*—was apparent when Suburb attempted to change the routine for following patients indefinitely. Research indicated it did not improve survival. Oncologists and one surgeon were initially opposed, but the surgeon leader and the general practitioner questioned current practice. In the case of following patients, the surgeon leader also felt others should decide while citing evidence women found most recurrences themselves.

Echoing the influence of hierarchy in many projects, some individuals were reluctant to challenge the surgeon leader's status. Yet, he increased the number of brokers negotiating a new practice. Even so, differences persisted, as they often do. Pathologists at Suburb, for example, continued to report findings in minute detail typical of their disciplinary training and practice. In contrast, surgeons often came to the pathology department to discuss expert opinions. Pathology at Suburb belonged traditionally to the university, so it had an academic focus. Read through the lens of communities of practice, Oborn and Dawson concluded, this case underscored the role of power. It also recalls to mind Choi and Richards' (2017) concept of epistemic rights in Chapter 4. Status flows to individuals whose knowledge is deemed most valued. Reflective of medical hierarchy, nurses did not participate actively and usually sat in back in the meeting room, while radiation oncologists tended to defer to surgeons. Spatial separation of participants is not trivial: it is indicative of a power hierarchy.

Toward Failure and Success

Two final questions follow from examples in this chapter. First, what is the nature of expertise for crossdisciplinary and cross-sector work? Collins and Evans (2009) distinguished two forms. *Interactional expertise* is sufficient familiarity to interact with other participants in a group, while *contributory expertise* is possession of enough knowledge of language and norms of other subcultures to participate fully. The latter capacity, however, does not assume mastery of another discipline. Enculturation in a community, Collins and Evans explained, transmits explicit rules of practice and tacit assumptions. However, adequacy is a more accurate characterization of what *contributory expertise* entails. Some concepts, vocabulary, data procedures, and other aspects of research can even become standardized among individuals in a collaborative project. The distinction between I- and T-based training also captures differences. In contrast to the columnar I-shape of single disciplinary or professional preparation, T-shaped training entails deep knowledge in the vertical trunk of one area and breadth in the horizontal bar. In addition, the concept of shield-shaped competency combines breadth with depth in multiple areas (Bosque-Pérez et al., 2016).

The second question follows from the first: What are the implications for career prospects? Catherine Lyall (2019) invoked Katrine Lindvig's (2017) distinction between loud and performative voices of interdisciplinarity. Loud voices emanate from pronouncements at institutional levels, while performative voices unfold in the quiet of daily practice. Despite shared enthusiasm for benefits among students and vice rectors of research in the UK, data from Lyall's semi-structured interviews were mixed. In the UK the Research Excellence Framework is a major disincentive, anchoring criteria of quality in discipline-based research. Students also had different motivations for doctoral work. One group was dedicated to pursuing interdisciplinary careers from the outset, influenced by prior education and training or having already identified pertinent research topics. A second group was sensitized via an interdisciplinary undergraduate degree, then influenced by supervisors. A third group, in turn, was opportunistic and also encouraged by supervisors. Aspirations differed as well. Some students wanted to stay in academia or become a university researcher. Others wanted to continue doing research but not necessarily do so in an academic role or setting.

Lyall situated responses from her informants, along with insights from literature review, within an institutional ecosystem that includes both funding and organizational parameters involving members of promotion committees, heads of departments, deans, and vice rectors as well as journal editors, grant reviewers, learned academies and professional groups, and staff of funding agencies. Individuals need to locate themselves within this ecosystem. Based on their experiences, the students she interviewed differed on whether interdisciplinarity has become more mainstream or is still limited by a particular discipline and unit. Lyall admonished, "Universities still need to find better ways of 'squaring the circle' of both incentivizing and then rewarding interdisciplinarity" (2019, p. 53). None of the students regretted becoming involved in interdisciplinary work. However, some expressed concern about finding a place, especially those hoping for academic careers. Talk of the next generation of inter- and trans-disciplinarians rings hollow when impediments continue to diminish future prospects. Discrepancies between perceptions of students and vice rectors in charge of research lead to the final chapter in this book, which addresses the topic of failure while offering final insights on conditions for success.

6
From Failing to Succeeding

> This study identified academic institutional customs that create a small but persistent "drag" on researchers who would like to do interdisciplinary research and teaching. They include especially the academic promotion and reward system and the department-based budgeting structures of universities.
>
> **NASEM (2005)**

> Studies in the higher education literature have identified an incongruity between the traditional university reward system and the "commercial ethos" that modern research universities encourage.
>
> **P. Craig Boardman and Branco L. Ponomariov (2007)**

The opening epigraphs underscore why after decades of recommendations from science-policy bodies, educational commissions, professional organizations, and lessons for best practices in growing numbers of case studies the same barriers and disincentives keep appearing . . . then reappearing. The first epigraph, from NASEM's 2005 state-of-the-art report on interdisciplinary research, states the matter succinctly. Despite widespread rhetoric of endorsement, the day-to-day reality in many academic institutions falls short of transforming academic values and structures, let alone bridging higher education and cross-sector partnerships. As a result, both existing initiatives and future possibilities are deterred. Reflecting on the state of the university in the 21st century, Crow and Debars (2017) concluded many institutions continue to lag behind in accommodating new forms of knowledge production, limiting their ability to address scientific and societal problems. The second epigraph highlights impediments to rewarding cross-sector work of an applied or commercial nature. Although Boardman and Ponomariov (2007) emphasized entrepreneurial work, their caveat extends to community engagement as well, minimizing and even outright penalizing careers of faculty who engage in related practices.

Beyond Interdisciplinarity. Julie Thompson Klein, Oxford University Press. © Oxford University Press 2021.
DOI: 10.1093/oso/9780197571149.003.0007

Both epigraphs also suggest failure is not straightforward. O'Rourke and Fam (2021) identified a number of reasons when introducing the first book devoted to the topic in crossdisciplinary work. They include setting unrealistic expectations and goals, becoming derailed by unexpected developments, not satisfying funders' requirements, and lacking adequate resources. In the same book, Martina Utkowitz (2021) observed the enigmatic nature of failure might mean a range of outcomes: from a total breakdown and falling short of an ideal to a mistake stemming from an accident or other misfortune or even a sense of disquiet. In another chapter Vilsmaier and Thalheimer (2021) noted failure might mean something was missing or insufficient or simply "didn't work well." Bellotti and D'Agostino (2021), in turn, cited shortfalls due to an inadequate number of crossdisciplinary researchers, ad hoc status, insufficient time and budget, absence of a strong common vision, and inadequate engagement of stakeholders in communities and policy circles. Additional reasons include lack of requisite skills, insufficient momentum, inadequate communication, uneven participation, conflict, disciplinary priorities, inadequate leadership, and turnover of personnel.

Failure, then, is a heterogeneous concept, and outcomes may range from complete collapse of a project to minor glitches. Roderick Lawrence further questioned criteria for determining failure given co-existence of multiple kinds of interdisciplinarity (pers. comm., 5 December 2015) And, Cynthia Mitchell replied in an online exchange that something deemed a failure at one point might lead to a better outcome in the long run (pers. comm., 7 December 2018). Instead of closing the book by compiling yet another litany of recommendations, this chapter begins by condensing barriers and impediments into a digest of challenges in both crossdisciplinary and cross-sector work. It then elaborates reasons for shortfalls with examples highlighting impediments to radical forms of interdisciplinarity and long-term viability. The chapter turns next to six overarching principles for success: transparency; informed use of best practices, models, guidelines, and reports; consistency and alignment of activities in a systematic approach; balance of disciplinary, professional, crossdisciplinary, and cross-sector work; credit for boundary crossing; and appropriate criteria of evaluation. This section also considers the role of technology, the academic reward system, and responsibility for change. The chapter and the book close by returning full circle to the question of what constitutes interdisciplinarity today, followed by five gateways into the burgeoning body of resources.

Failing

The place to start is asking whether common reasons for failing appear. Benchmark reports have identified an abundance of barriers and disincentives: including NASEM's *Facilitating Interdisciplinary Research* (2005) and *Convergence* (2014), the European Union Research Advisory Board's (EURAB) *Interdisciplinarity in Research* (2004), the League of European Research Universities' (LERU) *Social Sciences, Humanities, and Interdisciplinary Research* (2014), NASEM's *Enhancing the Effectiveness of Team Science* (2015), the British Academy's *Crossing Paths* (2016), the Global Research Council's *Interdisciplinarity: Survey Report for the Global Research Council 2016 Annual Meeting* (Gleed and Marchant, 2016), and the 2017 INTREPID series on bridging sciences and social sciences with humanities and arts. This chapter also draws insights from a new policy paper, *Addressing Societal Challenges Using Transdisciplinary Research* (OECD, 2020), with lessons from systematic analysis of methods and practices at the project level across countries. Although reports emanated from different contexts, they were all prompted by the gap between rhetoric of endorsement and realities of practice.

In the aggregate authoritative reports and other accounts yield a composite roster of factors, captured in Table 6.1. The table recognizes both crossdisciplinary and cross-sector factors, expanding my earlier digest focused on academic institutions (Klein, 2010). They are categorized in primarily academic factors and factors appearing both within and across sectors.

Despite sorting into separate categories, impediments are systemic. Organizational Structure and Administration influence decision-making, distribution of budgets, possibilities for and limits to change, and prospects for building and sustaining projects and programs. Procedures and Policies limit flexibility, impede autonomy, and enforce values and criteria of quality that do not fit complex needs of both crossdisciplinary and cross-sector work. Funding typically favors established forms of expertise in prescribed organizational slots, follows standard procedures, and perpetuates inadequate support for alternatives and experimentation. Resources and Infrastructure also tend to favor dominant disciplines and professions; are insufficient for boundary-crossing research and education; and do not support long-term follow-up, professional development, or making good on the oft-touted claim of educating the next generation of researchers and educators. Finally, Recognition and Reward reinscribe privileged values, lack sufficient criteria for evaluating both programs and individual achievements, and undermine

Table 6.1 Barriers and Disincentives of Crossdisciplinary and Cross-sector Boundary Crossing

Organizational Structure and Administration	
Primarily academic	• Discipline- and department-based silos • Territoriality and turf battles • Ambiguous status of projects, programs, and units within an institution • Insufficient control over budget, research, and curriculum
Within and across sectors	• No clear and authoritative report lines for dedicated programs and units • Rigid one-size-fits-all model of organizational and/or professional structure • Ad hoc and dispersed infrastructure • Lack of experienced leaders • Resistance to innovation and risk • Shortage of incubator space for working across boundaries
Procedures and Policies	
Primarily academic	• Inflexible rules for approval of new programs and courses • Unfavorable policies for allocating workload credit in teaching • Rigid and exclusionary degree requirements • Targeted programs and electives for particular cohorts rather than mainstream practice • Lack of inclusive guidelines for hiring, tenure and promotion, and salary • Unfavorable policies for intellectual property and indirect cost recovery from grants
Within and across sectors	• Prioritizing of some approaches over others • Narrow strictures of resource allocation • Dominance of and defaulting to disciplinary, professional, and sector perspectives • Competing rules across domains and sectors
Funding	
Primarily academic	• Budget allocations favoring discipline-based units • Inadequate funding for seeding new projects and programs • Dependence on soft money without hard budget allocation • Lack of access to overhead from grants
Within and across sectors	• Narrow criteria for funding and siloed categories • Shortage of sources for financing • Reliance on volunteerism and overload • Joint funding by multiple agencies/research councils with differing demands
Resources and Infrastructure	
Primarily academic	• Inadequate resources for activities, including space and equipment • Inadequate personnel for education and research, including faculty lines • Restricted access to internal incentives and seed funds for ID research and curriculum • Insufficient time to learn language and culture of another discipline • Insufficient time to form collaborative relationships in teams • Insufficient support for student assistantships and fellowships
Within and across sectors	• Weak or no professional development system • Ignorance of pertinent literature and networks • Insufficient time for planning and implementation of cross-boundary projects • Demands for evidence of impact on short timescales

Table 6.1 *Continued*

Organizational Structure and Administration	
Recognition and Reward	
Primarily academic	• Lack of support at department, college, and/or university levels • Invisibility and marginality of individuals and teams • Lack of appropriate criteria in tenure and promotion system • Ineligibility for awards, honors, incentives, and development programs
Within and across sectors	• Lack of long-term sustainability, rendering projects and programs ephemeral • Negative bias against cross-disciplinary, interprofessional, and/or cross-sector work • Weak networking channels and communication forums • Inadequate incentives for participation

Sources: Klein; Laursen Evaluation and Design, LLC, 2021.

willingness to take risks. A deeper look at several examples underscores why initiatives fall short despite original hopes while depicting different outcomes of interdisciplinary efforts.

Evans and Marvin's (2006) case study of research on urban sustainability in the UK during the 1990s underscores the challenge radical initiatives face. The first of three forms of interdisciplinarity posited a transformative goal of treating the city as an integrated combination of science, technology, and society. The second and third forms diminished that vision. The second parceled components into three separate councils, while the third outsourced responsibilities to local authorities in user communities. Failing to meet projected hopes for the first form will ring true for veterans of many projects. The first form did not succeed because practical problems proved more intractable than envisioned, common ground was smaller than assumed, and disciplinary identities remained intact. Evans and Marvin acknowledged the outcome might be interpreted as a failure of nerve and the power of disciplinary interests and loyalties. Only one of three attempts to understand the city in a radical way approached that goal. However, it too failed. Instead, they questioned prescriptive definitions of interdisciplinarity and the assumption an integrated account can synthesize diversity in a project focused on a complex problem. The track record of interdisciplinary programs is also marked by challenges of implementing and sustaining innovative approaches even in celebrated models.

Augsburg and Henry's (2009b) collection of institutional case studies, aptly entitled *The Politics of Interdisciplinary Studies*, asserted strong programs have a self-conscious spine of interdisciplinary courses. In contrast, weak programs lack core principles and clear processes of integration, relying instead on multidisciplinary distribution across discipline-based courses. Even strong programs, however, have been terminated or, in their words, declawed by reduced budgets and other resources, raised class sizes, and dispersal into other units. William Newell (2009) further cautioned interdisciplinary programs are always at risk. He likened them to low-hanging fruit pitted against disciplinary hegemony in a zero-sum game of competition for resources, enrollments, and ideologies. Disciplinary hegemony alone, however, does not account fully for demise of programs. Heightened priority of marketable majors coupled with devaluing of liberal arts has undermined the intellectual viability of many programs, privileging instrumental goals instead. In addition, increasing corporatization of the university accentuates financial exigencies that favor traditional disciplines. As a result, budget cutbacks often result in experimental programs succumbing to the syndrome of last hired–first fired.

Even lauded strategies do not guarantee success. The cluster hiring initiative, for example, facilitates multiple hires across a campus around a common theme, a strategy that in many cases has been a catalyst for new fields. Individuals are typically housed within an existing department, though they may also affiliate with a research center. Control of a position typically resides in the office of a top-level administrator, often the provost, so departments do not retain them. Not all clusters, however, have resulted in genuine collaboration, and departure of one or more appointees undercuts momentum for an integrated cluster. Kerri Holley (2017) reported interdisciplinary programs experience vulnerabilities regardless of funding mechanism. If freestanding units are funded directly through a central administration and general fund, they become reliant on administrators' willingness to continue supporting them. Departure of an administrative champion has jeopardized and even scuttled more than one program. If instead freestanding units are self-funded, they must rely on student tuition, state appropriations for public campuses, and/or external grants. Because enrollment is the measure by which a program is evaluated, low numbers also amplify vulnerability.

In presenting a cautionary tale of an environmental studies program, Wallace and Clark concluded barriers operate both independently and in concert to create an "organizational pathology" (2017, p. 236). Once again, veterans of other programs will recognize challenges that arose. The program emerged during a flurry of comparable programs propelled by the environmental movement of the late 1960s. Located in a private undergraduate liberal arts college in the US, it was originally housed in a geology department and touted as the school's first interdisciplinary major. After campus debate, the curriculum was approved and two tenured faculty in natural sciences stepped forward to serve as chairs. Two rotating faculty from natural sciences and social sciences taught the core courses, with additional electives across departments. Yet, despite some successes, the program was more multidisciplinary than interdisciplinary. Faculty work also relied on overload, a recipe for both low morale and burnout. The program faced other obstacles as well, including inflexible structuring of curricular requirements, the expectation of serving a core curriculum, replacement of the original tenured leaders with two tenure-track faculty, and reduction of oversight from a chair to a director given the untenured status of a subsequent program director. Other challenges followed as well.

When the two open environmental studies positions were consolidated into one senior-level tenured hire, after their occupants departed for other jobs, a non-tenure-track lecturer was appointed to direct it; then the candidate selected in a national search for a tenured director wound up favoring his own research interests over deep integration. Ultimately the number of faculty and courses shrank. Other programs, especially STEM-related ones, took advantage of the opportunity to claim the mantle of interdisciplinarity. Wallace and Clark's account affirms that without voice and power programs are marginalized. Top administrators need to understand and support interdisciplinarity, while faculty and staff must be proactive in developing related methods and content. Even a seemingly favorable institution, though, is no guarantee. Imbruce and Prazak (2021) also recalled the Mill Town Project's effort to introduce an environmental studies course at a small liberal arts university in the US focused on postindustrial life. Despite not having fixed departments or majors, discipline-based priorities lessened commitment to interdisciplinarity. Three years of conversations yielded learning about people in their environments among members of a crossdisciplinary project team. Yet, sciences enjoyed higher status. Interest in environmental concerns continues, with some coursework, but in an ad hoc manner.

Principles for Succeeding

Much of what may be considered failure, Frédéric Darbellay suggested, can be translated into conditions for success (pers. comm., 5 December 2018). Flipping failure on its head produces enabling factors. However, even though best practices set standards for strength and rigor, Bozeman and Boardman (2013) cautioned they may not work in all contexts. O'Rourke and Fam

(2021) also advised projects do not always follow step-by-step protocols and theories. Moreover, success in one context does not necessarily translate to another. Daniel Stokols highlighted the value of multi-scale systems thinking for identifying key contextual variables. Situated at multiple environmental and social scales, they influence a team's or an organization's amenability to trans-disciplinary collaboration and integration. One type of leveraging—Informational—entails the meaning of a concept. A second type—Motivational—spans both personal stakes and group dynamics as well as leaders who can "seed the system" with a transdisciplinary orientation. A third—Material Resources—extends beyond individuals to include funding, physical space, equipment, and communication tools. The fourth level of leveraging—Structural and Institutional—touches on all five categories of Table 6.1, including barriers arising from policies, practices, and cultures (pers. comm., 6 July 2019; elaborated in Stokols' 2018 book).

As previous chapters demonstrated, no two crossdisciplinary projects are identical. Yet, in the aggregate lessons from case studies combined with recommendations in authoritative reports and insights from the larger literature, generate six general principles:

1. Transparency in all aspects of boundary work;
2. Informed use of best practices, models, guidelines, and authoritative reports;
3. Consistency and alignment of separate activities in a systematic approach;
4. Effective balance of disciplinary, crossdisciplinary, and cross-sector work;
5. Credit for rather than marginalization or dismissal of boundary-crossing work;
6. Appropriate criteria and a multimethodological approach to evaluation.

The first principle affirms the centrality of communication. Many projects falter when participants realize they have common interests but differing stakes. The second reinforces the importance of learning from experience and pertinent literature. The third minimizes fragmentation, defaulting to a particular domain or, to borrow Ellen Messer-Davidow's (2002) characterization, a multidisciplinary mélange. The fourth underscores the need for appropriate mixes of expertise driven by a particular problem or question. The fifth emphasizes recognition and reward, while the sixth moves beyond discipline-based criteria and proxy measures inadequate for assessing the quality of crossdisciplinary and cross-sector work. The six principles are not magic bullets, however. No set of recommendations ever is. Echoing Evans and Marvin (2006), the typical pattern is accommodation rather than radical transformation. Comparably, Catherine Lyall (2019) observed, change is more often a recalibration than a radical reconceptualization and redesign of the academy. The 2004 EURAB report urged integrating centers into activities of traditional departments, and the 2016 British Academy report advocated individuals have a departmental location and work within the institutional grain.

Technology is another prominent topic in discussions of success, in two respects. First, authors of the NASEM's 2005 report, *Facilitating Interdisciplinary Research,* called new technologies a major driver. Sophisticated instrumentation and new tools of computer-assisted mathematical analysis are facilitating sharing large amounts of data across disciplinary boundaries in areas as diverse as bioinformatics, geosciences, digital humanities, new media arts, and cognitive science. The implications are not merely technical. Alan Leshner observed "new technologies are driving scientific advances as much as the other way around" (2004, p. 729). Terry Shinn's (2001) characterization of research technologies further captures their crossdisciplinary and cross-sector nature. They are being produced by interstitial communities rather than single institutional, disciplinary, or industrial domains. The devices they create are generic, not designed for a single need. Technologies are also generating new ways of representing events and phenomena, including tools of visualization. Furthermore, they are not limited to a single knowledge community. The second reason is increased use of Web 2.0 technologies that facilitate collaboration.

The 2015 NASEM report on team science described formats ranging from international collaboratories to tools facilitating communication (e.g., email, wikis, chat rooms, platforms for teleconferencing), coordination (e.g., shared calendars, scheduling software, visual displays, awareness tools), information repositories (e.g., databases, shared files, lab notebooks), and computational infrastructure (e.g., system architecture, networks, large-scale resources, human computation). Capacity for integrative team learning, Decuyper, Dochy, and Van den Bosschec (2010) also reported, includes the software of a team (e.g., non-material repositories such as individual memory, shared mental models, and a transactive memory system) as well as the hardware (e.g., databases, bulletin boards, expert systems, artifacts, and papers). Technology does not automatically guarantee success, however. It must be combined with social strategies that promote communication. Moreover, Cummings and Kiesler cautioned, technology has not "conquered distance" (2005, p. 704). Time zones and institutional calendars, including semester versus quarter terms, impede scheduling meetings. In addition, institutions have different missions and capacities for crossdisciplinary and cross-sector work.

Cummings and Kiesler's study of 62 scientific collaborations further indicated multi-university programs were not so well coordinated as and reported fewer positive outcomes than projects with Principal Investigators (PIs) from fewer universities and co-located teams. The most effective coordination mechanism was direct supervision by faculty and graduate students, but it was used more on single campuses than for projects spanning multiple institutions. Holding a workshop or conference and travel was somewhat effective in helping generate new ideas. Others also scheduled monthly telephone meetings and workshops, which improved outcomes. Yet, communication technology such as email, instant messages, phone conferences, and videoconferences did not give PIs at multiple universities an added advantage. Websites were common but rarely used for ongoing work. Furthermore, while open-source platforms hold promise for communication and collaboration, complications arise in sharing data, including different protocols for privacy, security, and ethics as well as regulatory restrictions. Cross-sector collaboration can also be impeded by the proprietary nature of data in industry and government versus transparency in sharing data and publishing findings in academic settings.

The academic reward system is another prominent topic. Individuals and provosts responding to a preliminary survey for NASEM's 2005 report on *Facilitating Interdisciplinary Research*, ranked promotion criteria the top impediment, and NASEM's 2014 report on *Enhancing the Effectiveness of Team Science* indicated most universities still lacked comprehensive and explicit criteria for evaluating individuals' contributions to team science. As a result, they face a double handicap. Their work is typically judged by discipline-based standards, and contributions to collaborative research are undervalued if not by a first author or a PI. Moreover, echoing Boardman and Ponomariov in the second epigraph of this chapter, applied and commercial work continue to be marginalized or discounted. Gabriele Bammer (2012) also urged considering limits of available funding, time, and human resources as well as differing methodologies and types of interdisciplinarity. Based on literature review, Klein and Falk-Krzesinski (2017) further proposed a framework for appropriate evaluation. Creating a culture of reward, they concluded, requires alignment and consistency at all points, from hiring to preparing dossiers and judging candidates. Individual contributions to teamwork need to be documented and the nature of the work and pertinent venues explained.

Appropriate criteria are needed as well. Jacobs and Frickel (2009) contended quality of funding decisions, in particular, entails knowing when to elevate one set of disciplinary priorities over others when assigning value to proposals that exhibit intellectual breadth. However, they sidestepped the question of what constitutes quality for projects that cross boundaries. A fuller range of indicators includes evidence of the interdisciplinary substance of the work, unanticipated outcomes, and long-term outcomes usually ignored in short-term project cycles (Klein, 2008). Authors of the draft of the policy paper from the Global Science Forum, *Addressing Societal Challenges Using Transdisciplinary Research* (OECD 2020), cited an expanded scope of research, incorporation of other approaches, more innovative and holistic solutions, integration of practical and experiential knowledge, and attention to norms, ethics, and abstract ideas in case-specific knowledge. In addition, social benefits include venues for dialogue and conflict resolution as well as new relationships among academics, citizens, and stakeholders. Studies of funding agencies and the reward system yield further insights into pertinent criteria.

The British Academy, in turn, urged considering the purpose of a particular project or program, awareness of relevant literature, and inclusion of an explicit plan for communication, collaboration, and integration. Funding agencies differ, however. Based on comparison of the European Research Council and the US-based National Science Foundation, König and Gorman (2017) identified variances in selection of reviewers, relationships to disciplinary communities, and funding streams. They concluded agencies operationalize definitions within their means and constraints. Furthermore, agencies differ on offering targeted opportunities, let alone whether crossdisciplinary work occurs in regular funding programs. Beyond funding, reward systems also differ. For example, there is no exact equivalent of the US system of promotion and tenure in the UK, though Lyall's (2019) interviews of graduate students indicated those contemplating academic careers were concerned about tenure and risk. The British Academy also found concerns in applied and policy research are discounted for tenured positions. In addition, the EURAB and Global Research Council reports never used the word *tenure* but acknowledged widespread concerns about unfavorable policies and inadequate rewards.

Local institutions vary in readiness, including both human and material resources as well as their history and culture of both crossdisciplinary and cross-sector efforts. The Council of Environmental Deans and Directors exhorted the first step in handling interdisciplinary careers is holistic assessment of existing structures and cultures (Pfirman, 2011). The question of who is responsible for easing barriers follows. Benchmark reports designate actions for individuals and teams, universities, professional associations, funding agencies, government, industry, and communities. However, all parties are responsible. Augsburg and Henry (2009a) recalled Michel de Certeau's (1984) contention that tactics can never overthrow strategies of disciplinary hegemony. Yet, they admonished, professionalism is key, with explicit attention to integration, theory, and methodology. Presenting evidence-based evaluation of pedagogy and student learning is also crucial, while heeding best practices. Drawing parallel lessons from women's studies, Robyn Wiegman (2002) explained viability depends on interdisciplinary cores with a rigor that does not derive from generalist distribution or disciplinary accumulation. Moreover, a single strategy does not change an institutional culture. In testimony for the British Academy report, the Royal Academy of Engineering advocated a portfolio of options combining top-down, large-scale projects and smaller grants that might engender bottom-up interdisciplinarity.

The Future of Interdisciplinarity?

Previous chapters affirmed the oft-beckoned revolution has not arrived. However, institutional changes are underway despite differing opinions on how much and how fast they are occurring. Introducing the updated edition of *The Oxford Handbook of Interdisciplinarity*, Robert Frodeman (2017) raised three overriding questions about interdisciplinarity today:

1. What is its place in the political economy of knowledge?
2. Will it become central to transformation of the 21st-century university?
3. Are other vocabularies signaling displacement of the academy from the center of knowledge production, such as accountability, impact, and relevance?

Russell, Wickson, and Carew (2008) raised parallel questions about transdisciplinarity based on three drivers of change. The first—the knowledge economy of products and services—involves partnerships with industry and government. The second—an environmental imperative that renders knowledge a tradable commodity—has been a major catalyst for cross-institutional and transnational connections. The third—an engaged population—increases demands for trans-sector public participation at a time when the authority of science is weakening. Contradictions and tensions arise across all three drivers: consolidation versus interconnection, commodification versus mutual learning, and public versus private forms of knowledge. Like Frodeman, Russell, Wickson, and Carew also endorsed a socially robust problem-oriented approach to knowledge production. They consider the future of the university at stake as well. They are not alone in this view. Peter Weingart observed that the university has "lost its monopoly" as the sole institutional locus of knowledge production (2010, p. 12), while Craig Calhoun (2017) reported funders today often seek research for problem solving from think tanks.

These developments, it should be added, are taking place against the larger backdrop of declining public support for and trust in higher education, short-term and credential-based programs, devaluing of the professoriate in an adjunct labor force, and new exigencies driven by the global COVID-19 pandemic, among other crises. If transdisciplinarity is to prosper in universities, Russell, Wickson, and Carew (2008) contended, collaboration and teamwork must be valued along with a culture of reward and a spirit of mutual responsibility. Yet, they cautioned, no single model will boost the economy, save the environment, and empower communities. Moreover, "Transdisciplinarity is a practice, not an institution, and the more flexible, adaptable and open it remains, the greater will be its contribution" (p. 470). As for the future of interdisciplinarity, Frodeman asked whether it had reached a peak, with displacement and decline on the horizon, while Machiel Keestra (2019) advocated placing greater weight on actionability. The shift Keestra endorsed is grounded in the realization knowledge is both valid from different disciplinary perspectives and relevant in a social context.

Stokols' (2018) conceptual framework for a science of transdisciplinary action research recognized collaboration is not solely academic. Action research, he explained, unfolds in three ways: among scholars of different disciplines; among researchers from multiple fields and community practitioners who represent different professional and lay perspectives; and among community organizations across local, state, national, and international levels. Forms of collaboration, though, still differ. Some prioritize scientific discovery; others translate scientific findings into interventions aimed at alleviating societal problems; and others yet integrate disciplinary and professional expertise in a community-wide effort to improve environmental, social, and health outcomes. Action must not be divorced from imagination, however, a danger in the current heightened discourse of problem solving. Keestra exhorted action must entail reflective equilibrium based on weighing options in a process that may be likened to triple-loop learning in Chapter 5. Keestra's call also reinforces the importance of humanities. Yet, Barry and Born (2013) cautioned, the subordination-service mode of interdisciplinarity values humanities and social sciences only for their use to science and technology rather than their full complexity. Others also extol humanities. The 2014 LERU report heralded the importance of humanities for understanding both culture and behavior when addressing challenges of climate change, technology, and health. They are also crucial for comprehending ethics, rights, religious traditions, and linguistic diversity.

The SHAPE-ID project on Shaping Interdisciplinary Practices in Europe, funded by the European HORIZON 2020 initiative, is bridging Arts, Humanities, and Social Sciences (AHSS) with Science, Technology, Engineering, Mathematics, and Medicine (STEMM). Joel Graf called bridging them a "special case" of inter- and transdisciplinarity (2019, p. 35). Authors of a policy brief on SHAPE-ID observed, "While some AHSS disciplines and research may fit comfortably within the problem solving mode, others may understand their research in terms of critical or transgressive approaches and motives, such as challenging current narratives or bringing historical or contextual perspectives to bear on a situation to enhance understanding." Yet, referring to inter- and trans-disciplinary research, "[t]he dominant emphasis on IDR/TDR for problem-solving in the policy literature does little to encourage AHSS involvement, particularly participation from Arts and Humanities researchers" (Vienni Baptista, Lyall, Ohlmeyer, Spaapen, Wallace, and Pohl, 2020, p. 4).

Older organizations are also being challenged to rethink their mandates. In a major example, Newell (2013) highlighted challenges to the traditional mission of the Association for Interdisciplinary Studies (AIS). It emerged in 1979 with the aim of advancing interdisciplinary education in the US by emphasizing integration, undergraduate curriculum, social sciences, and humanities, as well as a complementary relationship to disciplines. However, he acknowledged, in recent decades the primary locus of interdisciplinary activity and funding has shifted from teaching to research, from undergraduate to graduate levels, from humanities and so-called soft social sciences to natural sciences and medicine, from an individual to a team activity, and from the ivory tower to the real world, including participation of non(*sic*)academics. In light of these developments, Newell asked whether AIS should re-evaluate and expand its understanding of interdisciplinary process and theory. Transdisciplinarity and team science, in particular, require AIS to rethink exclusive reliance of interdisciplinarity on disciplines, understanding over application, and intellectual inquiry rather than political or social activity. Newell cautioned focus on "interdisciplinarity itself" might be lost (p. 37).

The premise of interdisciplinarity itself, however, is challenged by heterogeneity of the concept. What, then, is the nature of interdisciplinarity? Keestra (2019) contended increasing inclusion of extra-academic stakeholders in transdisciplinary research constitutes a fundamental challenge to traditional notions of integrative interdisciplinarity. Commitments to particular objectives need not be sacrificed, including both critique and traditional values of liberal arts. However, he exhorted, they must now be situated within the changing landscape of research and education. Furthermore, as the brief opening historical sketch in this book revealed, changes in definition and priorities are not new. Nor are tensions surrounding meaning and purpose. In the 1920s, the core term signified both integrative curricula and problem-oriented research. Subsequently curricula expanded to include not only thematic study of great ideas enshrined in a canon of texts but also historically grounded study of contemporary problems. Amelioration of problem-oriented research followed, focused on social challenges as well as national defense, aeronautics, and science-based areas of international economic competition. What prospects, then, lie ahead at this point in roughly one century of the concept's history?

Reflecting on future prospects, Frédéric Darbellay (2019) sketched four scenarios. The prospects range from accommodation and assimilation to transgression and transformation:

- Scenario 1 conserves the dominant system of disciplinarity. It would tolerate multi-, inter- and trans-disciplinary approaches as long as they are not disruptive of the system. Thus, it is neither transformational nor critical.
- Scenario 2 is a form of incremental or progressive change without radical interrogation of disciplinary foundations. Hence it too conserves the status quo, even with circulation and negotiation of ideas, theories, and methods.
- Scenario 3 is more radical, demanding redefinition of the disciplinary system or an alternative. It risks conflict and even warfare, with lack of communication as well as epistemological and institutional blockages thwarting change.
- Scenario 4 postulates a postdisciplinary vision prioritizing openness, co-operation, and co-production in an anti or adisciplinary epistemological stance. It is a militant critical discourse that disrupts and transgresses disciplinary logic.

Scenario 4 might be likened to postnormal science and to a transdisciplinary vision of Mode 2 knowledge production. However, Darbellay questioned how radical they actually are. Postdisciplinarity is not synonymous with many connotations of interdisciplinarity and forms of transdisciplinarity that challenge necessity of a disciplinary foundation of knowledge. Yet, given the slow and uneven pace of change in academic institutions, Scenario 4 is more utopic than real. A study of dedicated centers and programs led Diana Rhoten (2004) to ask whether interdisciplinary research is a trend or a transition, given piecemeal efforts, superficial use of the label, and lack of systemic implementation across administrative levels. Even today, Crow and Debars (2017) concluded, the filiopietism of venerating tradition encourages adherence to historical models even after their relevance or usefulness has diminished. Continuing barriers and disincentives led Frodeman to question the degree to which crossdisciplinary work consists only of one-offs. Lyall (2019) further admonished the paradox of promoting interdisciplinarity while impediments continue is a "manifest misalignment." She added, "Tempting anyone into an interdisciplinary career without providing an adequate safety net still seems, at best, ironic if not hypocritical and possibly even unethical" (pp. 4, 104).

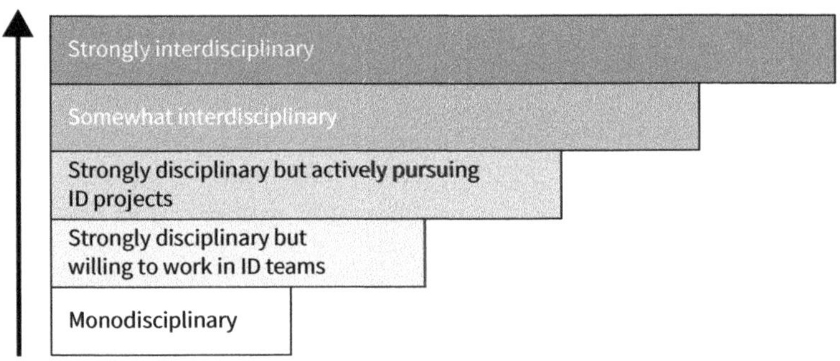

Fig. 6.1 Degree of interdisciplinary involvement.

Furthermore, all careers are not the same. The 2004 EURAB report declared neither training nor experience necessarily mean a person becomes interdisciplinary, while Lyall, Bruce, Tait, and Meagher (2011) identified a range of profiles depicted in Figure 6.1. The arrow in Figure 6.1 points upward, suggesting linear progression. However, Lyall, Bruce, Tait, and Meagher explained a person might work on a team at a particular point or conduct solo research or teaching. The degree and intensity of interdisciplinarity might also vary over the course of a career. Some may remain closely aligned with their disciplines, while others engage in academic forms of interdisciplinarity, and others yet concentrate on problems entailing interactions beyond the academy. Following suit, any individual may also embody one or more profiles. Darbellay (2019) described roles in multiple identities. A *Conservator* might evolve toward reform through crossdisciplinary dialogue or contest its destabilizing approach. A *Warrior* might retreat at some point to a moderate position or push for transformation. In contrast, a *Reformer* might work toward disruption and even become a Revolutionary. The latter two roles, moreover, exhibit typical traits of interdisciplinary individuals. He concluded "would-be agents of change must achieve legitimation of transgressive, nonconformist, and ultimately 'undisciplined' attitudes" (p. 106). Darbellay also noted a researcher's trajectory is built, evolves, transforms, and redefines itself in the long course of academic trajectories.

Based on a study of 66 researchers in 10 university structures, Darbellay (2017) identified multiple trends: a thematic identity, a hybrid identity, an interdisciplinary native, an interdisciplinary migrant, and an undisciplined identity. Individuals represented multiple fields since they were working in groups at crossroads of disciplines, such as environment and sustainable development, migration studies, health, neurosciences, and digital humanities. Thematic identity was prominent in fields examined in Chapter 3, including in this case environmental, migration, gender, and global studies. Researchers at crossroads of two or more disciplines also form a hybrid identity associated with interdisciplines, such as psycho-sociology and socio-anthropology. A native would likely be on an interdisciplinary trajectory from the beginning of education or training, without a fixed disciplinary anchor and with interests spanning a spectrum of disciplines. A migrant, in turn, is often confirmed in a discipline of origin but gradually opens to interdisciplinarity by borrowing concepts and circulating between disciplines. These identities, though, are not mutually exclusive. They can coexist and evolve.

Resources

Lyall, Bruce, Tait, and Meagher (2011) observed much of the knowledge surrounding capacity building for interdisciplinary work is tacit. Practitioners often learn by doing. Yet, failing to use pertinent literature and other resources is as inexcusable for crossdisciplinary and cross-sector work as it is for disciplines and professions. From the beginning of its history, multiple organizations and networks mentioned in prior chapters have presented strategies and models for change based on experience. The 2019 NASEM report of a workshop on fostering a culture of convergence, for example, proposed advisory teams of individuals with expertise in areas such as transdisciplinary research, team science, innovation studies, and organizational change. Provosts and senior research officers could aid in coordination, while directors of centers, programs, and projects could form peer groups to identify transferable elements across organizations. A broad-based convergence acceleration forum could share ongoing challenges and effective practices. Whether targeted to convergence or other special interests, networks and communities are also essential means of support for efforts at all levels and stages.

The References section of this book identifies many relevant works. However, five sites provide gateways into a greater abundance of resources:

Association for Interdisciplinary Studies (AIS), http://www.units.muohio.edu/aisorg/:

The AIS website provides original materials as well as the organization's newsletter and journal. The Resources link includes guidelines for tenure and promotion as well as assessing interdisciplinary general education and writing. Sublinks on the Publications page lead to an annotated list and an overview of definitions, philosophy, history, and best practices spanning communication, teaching, research, administration, and public policy analysis. In addition, the Reprints link includes a guide to syllabus preparation, a resource review, and an article on designing courses. The Interdisciplinary Connections includes groups with kindred interests.

Short Guides, https://www.wiki.ed.ac.uk/display/ISSTIInterdisciplinary:

The four-page digests cover a broad range of topics, including developing and reviewing research proposals, building and managing research teams, supervising doctoral students, and troubleshooting common challenges of management, as well as designing research for policy and practice and developing strategies for research groups. They also address funding, evaluation, leadership, and careers. The digests provide capsule materials for use in both education and professional development. In addition, the Guides, which emanated from interdisciplinary master classes, contain updated information about other projects and publications. They are downloadable for free from the wiki site.

i2S (Integration and Implementation Sciences), i2s.anu.edu.au

This Australian-based network aims to improve research on complex real-world problems while providing a common identity for researchers across domains. The Resources link presents tools, approaches, and cases, along with information about journals, professional associations, other networks, and conferences. The blog is a forum for sharing concepts and methods, with a short précis for training, education, and projects and programs. And *i2S News* provides updates. Resources are organized into three categories: synthesizing disciplinary and stakeholder knowledge, understanding and managing unknowns, and providing integrated support for changes in practice and policy.

Network for Transdisciplinary Research (td-net), http://www.transdisciplinarity.ch/

This trilingual website is a forum for news, definitions, and updates on conferences and a toolbox for co-producing knowledge in collaboration with stakeholders in society to address complex challenges. The toolbox covers knowledge synthesis and integration, stakeholder engagement, participatory research, team research, disciplinary collaboration, design thinking, and impact-oriented research. Users can link to selected methods with answers to questions, definitions, purposes, and strategies. The Publications link provides updates on literature, including reference works such as *The Handbook of Transdisciplinary Research* and *Principles of Transdisciplinary Research.*

International Network for the Science of Team Science (INSciTS), https://www.inscits.org

InSciTS is a home for team science researchers, educators, and others advancing the field. The SciTS and Team Science Resources link leads to selected references, kindred organizations, and information from annual meetings. Members of the network have also produced gateways: a Team Science Toolkit, a parallel Mendeley forum, and the Coalesce learning modules. Although currently dormant due to website redesign, the Toolkit's primary categories are tools and measures, supported by an annotated bibliography. Mendeley is an updated forum for information, and the first module in the Coalese compilation provides a generic introduction. (http://www.teamsciencetoolkit.cancer.gov/public/home.aspx; https://www.mendeley.com/community/science-of-team-science; https://www.teamscience.net)

Gateways have the added advantage of not only providing their own in-house resources but lead outward to others, expanding the wisdom of practice. Authors involved in creating the Short Guides, for example, published Lyall, Bruce, Tait, and Meagher's (2011) separate primer, *Interdisciplinary Research Journeys: Practical Strategies for Capturing Creativity.* It features case studies, key advice boxes, and checklists of questions on careers, project design, student supervision, assessment and program review. Members of INSciTS also produced the field guide *Collaboration and Team Science* (Bennett, Gadlin, and Marchand, 2018). It provides pragmatic strategies for building a team climate, with chapters focusing on leadership, trust, vision, communication, and credit as well as sharing, managing difference and conflict, sustaining and strengthening a team, and navigating and leveraging networks and systems. The site contains checklists for determining whether a particular approach is working, and the authors recommended other resources, such as Ledford's (2008) concept of a "collaborators pre-nup."

As the five sites also indicate, organizations sometimes interact to support crossdisciplinary and cross-sector work. A recently formed ITD Alliance aims to improve dialogue across key groups at an international level. Its founding members include key figures in AIS, td-net, INSciTS, and i2S. The Alliance lacks funding for long-term support and joint projects. Yet, individuals involved in the five gateways share common interests even with their different organizational missions (http://itd-alliance.org). Their efforts also underscore three final overriding lessons of this book. First, there is no one-size-fits-all. Practices and models in one location may not work in another, and generalizations about *the* nature of inter- or trans-disciplinary work must always be contextualized. Second, there are no magic bullets. Even tested strategies and models may not guarantee success, be sustainable, or work in a particular location. Third, informed awareness, particularly clusters of options and resources, is crucial. The challenge now is to make better use of them.

References

Abbott, A. (1988). *The system of professions: An essay on the division of expert labor.* Chicago: University of Chicago Press.
Aboelela, S., et al. (2007). Building a community of research practice: Intragroup team social dynamics in interdisciplinary mixed methods. *Journal of Mixed Methods Research, 7*(3), 261–273.
Addelson, K. P. & Potter, E. (1991). Making knowledge. In J. Hartman & E. Messer- Davidow (Eds.), *(En)Gendering knowledge: Feminists in academe* (pp. 259–277). Knoxville: University of Tennessee Press.
Akkerman, S. & Bakker, A. (2011). Boundary crossing and boundary objects. *Review of Educational Research, 81*(2), 132–169.
Albert, M, Laberge, S., & Hodges, B. D. (2009). Boundary-work in the health research field: Biomedical and clinician scientists' perceptions of social science research. *Minerva 47,* 171–194.
AMA Staff. (2019). AMA Critical Skills Survey: Workers need higher level skills to succeed in the 21st century. AMA, 24 January. https://www.amanet.org/articles/ama-critical-skills-survey-workers-need-higher-level-skills-to-succeed-in-the-21st-century/.
American Academy of Arts and Sciences. (2013). *Arise II: Unleashing America's research and innovative enterprise.* May. http://www.amacad.org//publication/arise-ii-unleashing-americas-research-innovation-enterprise.
Andersen, H. & Wagenknecht, S. (2013). Epistemic dependence in interdisciplinary groups. *Synthese, 190*(11), 1881–1898.
Apostel, L., et al. (Ed.). (1972). *Interdisciplinarity: Problems of teaching and research in universities.* Paris: Organization for Economic Cooperation and Development.
Argyris, C. & Schön, D. (1996). *Organizational learning II: Theory, method and practice.* Reading, MA: Addison-Wesley.
Armstrong, C. & Fontaine, S. L. (1989). The power of naming: Names that create and define the discipline. *Writing Program Administration, 13*(1–2), 5–14.
Arnstein, S. R. (1969). A ladder of citizen participation. *Journal of the American Planning Association, 35*(4), 216–224.
Arocena, R. & Sutz, J. (2016). Development studies as an interdisciplinary field: Research, teaching, and institution building in Uruguay. *Issues in Interdisciplinary Studies, 34,* 164–182.
Augsburg, T. (2014). Becoming transdisciplinary: The emergence of the transdisciplinary individual. *World Futures, 70*(3–4), 233–247.
Augsburg, T., & Henry, S. (2009a). Introduction. In T. Augsburg & S. Henry (Eds.), *The politics of interdisciplinary studies: Essays on transformations in American undergraduate programs* (pp. 7–27). Jefferson, NC: McFarland.
Augsburg, T., & Henry, S. (Eds.). (2009b). *The politics of interdisciplinary studies: Essays on transformations in American undergraduate programs.* Jefferson, NC: McFarland.
Auslander, P. (Ed.) (2003). *Performance: Critical concepts in literary and cultural studies.* London and New York: Routledge,
Baker, M. J. (2015). Collaboration in collaborative learning. *Interaction Studies, 16*(3), 451–473.
Bakhtin, M. (1981). The dialogic imagination: Four essays. Translated by C. Emerson M. Holquist. Austin and London: University of Texas Press.

Bammer, G. (2012). *Strengthening interdisciplinary research: What it is, what it does, how it does it, and how it is supported.* Canberra: Report for the Australian Council of Learned Societies (Melbourne).

Barry, A. & Born, G. (2013). Interdisciplinarity. In A. Berry & G. Born (Eds.), *Interdisciplinarity: Reconfigurations of the social and natural sciences* (pp. 1–56). London and New York: Routledge.

Barthes, Roland. (1977). From work to text. In *Image, music, text* (S. Heath, Trans.) (pp. 155–164). New York: Hill and Wang, Collins.

Bass, L. W. (1975). *Management by task forces: A manual on the operation of interdisciplinary teams.* Mt. Airy, MD: Lomond.

Bathrick, D. (1992). Cultural studies. In J. Gibaldi (Ed.), *Introduction to scholarship in modern languages and literatures* (pp. 320–340). New York: Modern Language Association of America.

Baxi, U. (2000). Transdisciplinarity and transformative praxis. In M. Somerville & D. Rapport (Eds.), *Transdisciplinarity: Recreating integrated knowledge* (pp. 77–85). Oxford: EOLSS.

Beauchamp, C. & Beauchamp, M. H. (2013). Boundary as bridge: An analysis of the educational neuroscience literature from a boundary perspective. *Educational Psychology Review, 25*, 47–67.

Becher, T. (1990). *Academic tribes and territories: Intellectual enquiry and the culture of disciplines.* Philadelphia: Open University Press.

Becher, T. & Trowler, P. R. (2001). *Academic tribes and territories: Intellectual enquiry and the culture of disciplines* (2nd edition). Philadelphia: Open University Press.

Beddoes, K. (2014). Methodology discourses as boundary work in the construction of engineering education. *Social Studies of Science, 44*(2), 293–312.

Bellotti, B. & D'Agostino, F. (2021). A fragile existence: A transdisciplinary food systems research program cut short. In D. Fam & M. O'Rourke (Eds.), *Interdisciplinary and transdisciplinary failures: Lessons learned from cautionary tales* (pp. 62–77). London and Routledge.

Bellotti, W. (2017). Transdisciplinarity as an emergent property in an agricultural research-for-development project on the East India Plateau. In D. Fam, J. Palmer, C. Riedy, & C. Mitchell (Eds.), *Transdisciplinary research and practice for sustainability outcomes* (pp. 221–235). London and New York: Routledge.

Bench, H. (2018). Arranging (*Enchaînement*). In C. Lury et al. (Eds.), *Routledge handbook of interdisciplinary research methods* (pp. 41–46). London and New York: Routledge.

Bender, T. & Schorske, C. S. (1997). Introduction. In T. Bender & C. S. Schorske, (Eds.), *American academic culture in transformation: Fifty years, four disciplines* (pp. 3–13). Princeton, NJ: Princeton University Press.

Bennett, D. (1997). Innovation in the liberal arts and sciences. In R. Orrill (Ed.), *Education and democracy: Re-imagining liberal learning in America* (pp. 131–149). New York: College Board.

Bennett, L. M., Gadlin, H., & Marchand, C. (2018). *Collaboration and team science: A field guide* (2nd edition). Bethesda, MD: Research Strategy, National Cancer Institute, National Institutes of Health. NIH Publication No. 18-7660.

Bergmann, M., Jahn, T., Knobloch, T., Krohn, W., Pohl, C., & Schramm, E. (2010). *Methoden Transdisziplinärer Forschung: Ein Uberblick mit Anwendungsbeispielen.* Frankfurt/Main: Campus Verlag.

Bergmann, M., Jahn, T., Knobloch, T., Krohn, W., Pohl, C., & Schramm, E. (2012). *Transdisciplinary research methods.* Chicago: University of Chicago Press.

Bird, E. (2001). Disciplining the interdisciplinary: Radicalism and the academic curriculum. *British Journal of Sociology of Education Studies, 30*(3), 509–533.

Blume, L. B. (2014). Making connections: Toward a transdisciplinary family science. *Journal of Family Theory and Review, 61*,1–4.

Boardman, P. C. & Ponomariov, B. L. (2007). Reward systems and NSF university research centers: The impact of tenure on university scientists' valuation of applied and commercially relevant research. *Journal of Higher Education*, *78*(1), 51–70.

Boix-Mansilla, V. (2017). Interdisciplinary learning: A cognitive-epistemological foundation. In R. Frodeman, J. T. Klein, & R. Pacheco (Eds.), *The Oxford handbook of interdisciplinarity* (pp. 261–275). Oxford: Oxford University Press.

Boix-Mansilla, V., Lamont, M., & Sato, K. (2016). Shared cognitive-emotional-interactional platforms: Markers and conditions for successful interdisciplinary collaborations. *Science, Technology, and Human Values*, *41*(4), 571–612.

Boix-Mansilla, V. & Lenoir, Y. (2010). Interdisciplinarity in U.S. schools: Past, present, and future. *Issues in Integrative Studies 28*, 1–28.

Borrego. M. & Newswander, L. K. (2010). Definitions of interdisciplinary research: Toward graduate-level interdisciplinary learning outcomes. *Review of Higher Education*, *34*(1), 61–84.

Bosque-Pérez, N., et al. (2016). A pedagogical model for team-based, problem-focused interdisciplinary doctoral education. *BioScience*, *66*(6), 477–488.

Boutet, D. (1993). Interdisciplinarity in the arts. *Harbour*, *6*, 66–72.

Bowker, G. & Star, S. (1999). *Sorting things out: Classification and its consequences*. Cambridge, MA: MIT Press.

Boxer, M. J. (1982). For and about women: The theory and practice of women's studies in the United States. *Signs*, *7*(3), 661–695.

Boyer, E. L. (1990). *Scholarship reconsidered: Priorities of the professoriate*. Princeton, NJ: Carnegie Foundation for the Advancement of Teaching.

Bozeman, B. & Boardman, C. (2013). *Research collaboration and team science: A state-of-the-art review and agenda*. Cham, Switzerland and New York: Springer.

Bozeman, B. & Youtie, J. (2017). *The strength in numbers: The new science of team science*. Princeton, NJ: Princeton University Press.

Bracken, L. J. & Oughton E. A. (2006). "What do you mean?" The importance of language in developing interdisciplinary research. *Transactions of the Institute of British Geographers*, *31*, 371–382.

Briggle, A. & Christians, C. G. (2017). Media and communication. In R. Frodeman, J. T. Klein, & R. Pacheco (Eds.), *The Oxford handbook of interdisciplinarity* (pp. 201–213). Oxford: Oxford University Press.

Brint, S. (2005). Creating the future: "New directions" in American research universities. *Minerva*, *43*, 23–50.

Brint, S. G., Turk-Bicakci, L., Proctor, K., & Murphy, S. P. (2009). Expanding the social frame of knowledge: Interdisciplinary degree-granting fields in American colleges and universities, 1975–2000. *Review of Higher Education*, *32*(2), 155–183.

Brister, E. (2016). Disciplinary capture and epistemological obstacles to interdisciplinary research: Lessons from Central African conservation disputes. *Studies in History and Philosophy of Biological and Biomedical Sciences*, *56*, 82–91.

Brown-Luthango, M. (2013). Community-university engagement: The Philippi CityLab in Cape Town and the challenge of collaboration across boundaries. *Higher Education: The International Journal of Higher Education and Educational Planning*, *65*(3), 309–324.

Brozek, J. & Keys, A. (1944). General aspects of interdisciplinary research in experimental biology. *Science*, *100*, 507–512.

Burtis, J. O. & Turman, P. D. (2006). *Group communication pitfalls: Overcoming barriers to an effective group experience*. Thousand Oaks, CA: Sage.

Brigandt, I. (2010). Beyond reduction and pluralism: Toward an epistemology of explanatory integration in biology. *Erkenntnis: An International Journal of Scientific Philosophy*, *73*, 295–311.

Briggs, C. (2008). Disciplining folkloristics. *Journal of Folklore Research: An International Journal of Folklore and Ethnomusicology, 45*(1), 91–105.

British Academy. (2016). *Crossing paths: Interdisciplinary institutes, careers, education, and applications.* London: British Academy.

Brown, J. S., Collins, A. & Duguid, P. (1989). Situated cognition and the culture of learning. *Educational Researcher, 18*(1), 32–42.

Bruce, A., Lyall, C., Tait, J., & Williams, R. (2004). Interdisciplinary integration in Europe: The case of the Fifth Framework Programme. *Futures, 36*(4), 457–470.

Burggren, W., Chapman, K., Keller, B., Monticino, M., & Torday, J. (2017). Interdisciplinarity in the biological sciences. In R. Frodeman, J. T. Klein, & R. Pacheco (Eds.), *The Oxford handbook of interdisciplinarity* (pp. 101–113). Oxford: Oxford University Press.

Burke, K. (1966). *Language as symbolic action: Essays on life, literature, and method.* Berkeley: University of California Press.

Burtis, J. O. & Turman, P. D. (2006). *Group communication pitfalls: Overcoming barriers to an effective group experience.* Thousand Oaks, CA: Sage.

Caldwell, L. K. (1983). Environmental studies: Discipline or metadiscipline? *Environmental Professional, 5*, 247–259.

Calhoun, C. (2017). Interdisciplinarity in the social sciences: Area studies, quantitative methods, and problem-oriented research. In R. Frodeman, J. T. Klein, & R. Pacheco (Eds.), *The Oxford handbook of interdisciplinarity* (pp. 117–130). Oxford: Oxford University Press.

Camic, C. & Joas, H. (2004). The dialogical turn. In C. Camic and H. Joas (Eds.), *The dialogical turn: New roles for sociology in the postdisciplinary age* (pp. 1–19). Lanham, MD: Rowman and Littlefield.

Campbell, N. & Kean, A. (1997). *American cultural studies: An introduction to American culture.* London and New York: Routledge.

Carlile, P. R. (2002). A pragmatic view of knowledge and boundaries: Boundary objects in new product development. *Organization Science, 13*(4), 442–455.

Carpenter, R. (2009). Boundary negotiations: Electronic environments as interface. *Computers and Composition, 26*, 138–148.

Carroll, Lewis. (1896). *Through the looking glass.* London: Macmillan.

Centellas, K. M., Smardon, R. E. & Fifield, S. (2014). Calibrating translational cancer research: Collaboration without consensus in interdisciplinary laboratory meetings. *Science, Technology, and Human Values, 39*(30), 311–335.

Center for Educational Research and Innovation. (1982). *The university and the community: The problems of changing relationships.* Paris: Organization for Economic Cooperation and Development.

Cetina, K. K. (2003). *Epistemic cultures: How the sciences make knowledge.* Cambridge, MA: Harvard University Press.

Chilisa, B. (2017). Decolonising transdisciplinary research approaches: An African perspective for enhancing knowledge integration in sustainability science. *Sustainability Science*, (5), 813+.

Choi, S. & Richards, K. (2017). *Interdisciplinary discourse: Communicating across disciplines.* London: Palgrave Macmillan.

Chou, W. H. & Wong, J.-J. (2015). From a disciplinary to an interdisciplinary design research: Developing an integrative approach for design. *International Journal of Art and Design Education, 34*(2), 206–223.

Ciccorio, R. A. (1970). Integration in the curriculum: An historical and semantic inquiry. *Main Currents, 27*, 60–62.

Clark, B. R. (1995). *Places of inquiry: Research and advanced education in modern universities.* Berkeley: University of California Press.

Clark, W. C., et al. (2011). Boundary work for sustainable development: Natural resource management at the Consultative Group on International Agricultural Research (CIGAR). Proceedings of the National Academy of Sciences. Online.

Collins, H. & Evans, R. (2009). *Rethinking expertise*. Chicago: University of Chicago Press.

Cornwall, G. & Stoddard, E. (1999). *Globalizing knowledge: Connecting international and intercultural studies*. Washington, DC: Association of American Colleges and Universities.

Costanza, R. (1990). Escaping the overspecialization trap. In M. E. Clark & S. A. Wawrytko (Eds.), *Rethinking the curriculum: Toward an integrated interdisciplinary college education* (pp. 95–106). New York: Greenwood.

Coyner, S. (1991). Women's studies. *NWSA Journal, 3*(3), 349–354.

Crenshaw, K. (1989). Demarginalizing the intersection of race and sex: A Black feminist critique of antidiscrimination doctrine, feminist theory, and antiracist politics. *University of Chicago Legal Forum, 5*, 139–168.

Cronin, G. (2014). Transdisciplinary scholarship: Integrating Boyer's model of scholarship with transdisciplinary research. *Transdisciplinary Journal of Engineering and Science, 5*(1): 14–20.

Crow, M. & Debars, W. (2017). Interdisciplinarity and the institutional context of knowledge in the American research university. In R. Frodeman, J. T. Klein, & R. Pacheco (Eds.), *The Oxford handbook of interdisciplinarity* (pp. 471–484). Oxford: Oxford University Press.

Crowley, S., Eigenbrode, S., O'Rourke, M., & Wulfhorst, J. D. (2014). Introduction. In M. O'Rourke, S. Crowley, S. Eigenbrode, & J. D. Wulfhorst (Eds.), *Enhancing communication and collaboration in interdisciplinary research* (pp. 1–10). Thousand Oaks, CA: Sage.

Cummings, J. N. & Kiesler, S. (2005). Collaborative research across disciplinary and organizational boundaries. *Social Studies of Science, 35*(5), 703–722.

Dallimer, M. & Strange, N. (2015). Why socio-political borders and boundaries matter in conservation. *Trends in Ecology and Evolution, 30*(3), 132–139.

Darbellay, F. (2015). Rethinking inter- and transdisciplinarity, undisciplined knowledge and the emergence of a new thought style. *Futures, 65*, 163–174.

Darbellay, F. (2017). From monomyth to interdisciplinary creative polymathy: The researcher with a thousand faces. *Journal of Genius & Eminence, 2*(2), 45–54.

Darbellay, F. (2019). From interdisciplinarity to postdisciplinarity: Extending Klein's thinking into the future of the university. *Issues in Interdisciplinary Studies, 37*(2), 90–109.

Darbellay, F., Moody, Z., & Lubart, T. (2017). Introduction: Thinking creativity, design, and interdisciplinarity in a changing world. In F. Darbellay, Z. Moody, & T. Lubart (Eds.), *Creativity, design thinking, and interdisciplinarity* (pp. xi–xxii). Cham, Switzerland and New York: Springer.

Darbellay, F., Moody, Z., Sedooka, A., & Steffen, G. (2014). Interdisciplinary research boosted by serendipity. *Creativity Research Journal, 26*(1), 1–10.

Darden, L. & Maull, N. (1977). Interfield theories. *Philosophy of Science, 44*(1), 43–64.

Davidow, E. (2002). *Disciplining feminism: From social activism to academic discourse*. Durham, NC: Duke University Press.

Davidson, C. N. & Goldberg, D. T. (2010). *The future of thinking: Learning institutions in a digital age*. London: MIT Press.

Davidson, C. N. & Janz, B. (2019). Theory into practice: Julie Klein's boundary work and institutional change. *Issues in Interdisciplinary Studies, 37*(2), 130–142.

Davies, J. (2011). Disciplining the disciplines. In P. Dawid, W. Twining, & M. Vasilaki (Eds.), *Evidence, inference, and equity* (pp. 37–72). Proceedings of the British Academy No. 171. London: Oxford University Press.

Davis, M. (2004). Bridging the gap or crossing a bridge? Indigenous knowledge and the language of law and policy. Paper presented to Bridging Scales and Epistemologies conference, Bibliotheca Alexandrina, Alexandria, Egypt, 17–20 March.

De Certeau, M. (1984). *The practice of everyday life*. Minneapolis: University of Minnesota Press.

Decuyper, S., Dochy, F., & Van den Bossche, P. (2010). Grasping the dynamic complexity of team learning: An integrative model for effective team learning in organizations. *Educational Research Review*, *5*(2), 111–133.

DeMelo, A. T. (2018). Abducting. In C. Lury et al. (Eds.), *Routledge handbook of interdisciplinary research methods* (pp. 90–94). London and New York: Routledge.

Derry, S., DuRussel, L. A., & O'Donnell, A. M. (1997). *Individual and distributed cognitions in interdisciplinary teamwork: A developing case study and emerging theory*. Research Monograph No. 7. Washington, DC: National Science Foundation.

Després, C., Fortin, A., Joerin, F., Vachon, G., Gatti, E., & Moretti, G. (2008). Retrofitting postwar suburbs: A collaborative design process. In G. Hirsch Hadorn et al. (Eds.), *Handbook of transdisciplinary research* (pp. 327–341). Cham, Switzerland and New York: Springer.

DeWachter, M. (1982). Interdisciplinary bioethics: But where do we start? A reflection on epochè as method. *Journal of Medicine and Philosphy*, *7*(3), 275–287.

DeZure, D. (2017). Interdisciplinary pedagogies in higher education. In R. Frodeman, J. T. Klein, & R. Pacheco (Eds.), *The Oxford handbook of interdisciplinarity* (pp. 555–572). Oxford: Oxford University Press.

Dielman, H. (2015). Transdisciplinary hermeneutics: Working from the inner self, creating ecologies of knowing. *Transdisciplinary Journal of Engineering and Science*, *6*, 72–85.

Dogan, M. & Pahre, R. (1990). *Creative marginality: Innovation at the intersections of social sciences*. Boulder, CO: Westview Press.

Dölling, I. & Hark, S. (2000). She who speaks shadow speaks truth: Transdisciplinarity in women's studies. *Signs*, *25*(4), 1195–1198.

Donadey, A. & Lionnet, F. (2007). Feminisms, genders, sexualities. In D. Nicholls (Ed.), *Introduction to scholarship in modern languages and literatures* (3rd edition) (pp. 225–244). New York: Modern Language Association.

Doucet, I. & Janssens, N. (2011). Editorial: Transdisciplinarity, the hybridization of knowledge production and space-related research. In I. Doucet & N. Janssens (Eds.), *Transdisciplinary knowledge production in architecture and urbanism: Towards hybrid modes of inquiry* (pp. 1–14). Cham, Switzerland and New York: Springer.

Downey, L. (2018). Figurationing. In C. Lury et al. (Eds.), *Routledge handbook of interdisciplinary research methods* (pp. 112–115). London and New York: Routledge.

Dressman, M., McCarthey, S., & Prior, P. (2009). Editors' introduction: Literate practices: Theory, method, and disciplinary boundary work. *Research in the Teaching of English*, *44*(2), 133–135.

Du Plessis, H., Sehume, J., & Martin, L. (2013). *The concept and application of transdisciplinarity in intellectual discourse and research*. Johannesburg: Real African Publishers.

DuRussell, L. A. & Derry, S. J. (2005). Schema (mis)alignment in interdisciplinary teamwork. In S. Derry, C. D. Schunn, & M. A. Gernsbacher (Eds.), *Interdisciplinary collaboration: An emerging cognitive science* (pp. 187–220). Mahwah, NJ: Earlbaum.

Eden, S., Donaldson, A., & Walker, G. (2006). Green groups and grey areas: Scientific boundary-work, nongovernmental organizations, and environmental knowledge. *Environment and Planning A*, *38*, 1061–1076.

Edwards, R. & Fowler, Z. (2007). Unsettling boundaries in making a space for research. *British Educational Research Journal*, *33*, 107–123.

Eigenbrode, S., et al. (2017). *Leading large transdisciplinary projects addressing social-ecological systems: A primer for project directors*. USDA NIFA. https://nifa.usda.gov/leading-transdisciplinary-projects.

Elam, D. (1990). Ms. en abyme: Deconstruction and feminism. *Social Epistemology*, *4*(3), 293–308.

Elgin, C. Z. (1996). *Considered judgment*. Princeton, NY: Princeton University Press.

Engbers, M. (2018). Kultur und differenz in der transdisziplinären nachhaltigkeitsforschung: analysen und konzeptionelle beiträge zur gestaltung von lern-/forschungsprojekten. Leuphana University,,PhD dissertation.

European Union Research Advisory Board. (2004). *Interdisciplinarity in research*. April. eurab_04_009_interdisciplinarity_research_final.pdf.

Evans, R. & Marvin, S. (2006). Researching the sustainable city: Three modes of interdisciplinarity. *Environment and Planning A, 38*, 1009–1028.

Fam, D., Lopes, A. M., & Mitchell, C. (2021). Transdiscplinary learning within tertiary institutions: A space to skin your knees. In D. Fam & M. O'Rourke (Eds.), *Interdisciplinary and transdisciplinary failures: Lessons learned from tales* (pp. 198–216). London and New York: Routledge.

Fam, D., Smith, T., & Cordell, D. (2017). Being a transdisciplinary researcher: Skills and dispositions fostering competence in transdisciplinary research and practice. In D. Fam, J. Palmer, C. Riedy, & C. Mitchell (Eds.), *Transdisciplinary research and practice for sustainability outcomes* (pp. 62–77). London and New York: Routledge.

Fam, D. & Sofoulis, Z. (2017). A "knowledge ecologies" analysis of co-designing water and sanitation services in Alaska. *Science and Engineering Ethics, 23*, 1059–1083.

Felt, U., Ingelsböck, J., Schikowitz, A., & Völker, T. (2013). Growing into what? The (un)disciplined socialization of early stage researchers in transdisciplinary research. *Higher Education, 65*, 511–524.

Fensham, R. & Heller-Nicholas, A. (2018). Making and assembling: Towards a conjectural paradigm for interdisciplinary research. In C. Lury et al. (Eds.), *Routledge handbook of interdisciplinary research methods* (pp. 29–40). London and New York: Routledge.

Fenske, M. (2007). Interdisciplinary terrains of performance studies. *Text and Performance Quarterly, 27*(4), 351–368,

Fernandes, V. & Phillippi Jr., A. (2017). Sustainability science: Political and epistemological approaches. In R. Frodeman, J.T. Klein, & R. Pacheco (Eds.) *The Oxford Handbook of Interdisciplinarity* (pp. 370–382). Oxford and New York: Oxford University Press.

Fiore, S. (2008). Interdisciplinarity as teamwork: How the science of teams can inform team science. *Small Group Research, 39*, 251–277.

Fiore, S. M. (2018). Better together. *Science, 359*(6376), 645.

Fiore, S. M., Gabelica, C., Wiltshire, T. J., & Sokols, D. (2019). Training to be a (team) scientist. In K. L. Hall, A. Vogel, & R. Croyle (Eds.), *Strategies for team science success: Handbook of evidence-based principles for cross-disciplinary science and practical lessons learned from health researchers* (pp. 421–444). Cham, Switzerland and New York: Springer.

Fish, S. (1989). Being interdisciplinary is so very hard to do. In *Profession 89*: 15–22.

Fisher, D. (1993). *Fundamental development of the social sciences: Rockefeller philanthropy and the United States Social Science Research Council*. Ann Arbor: University of Michigan Press.

Fleck, L. (1979). *Genesis and development of a scientific fact*. Chicago: University of Chicago Press.

Foucault, M. (1995). *Discipline and punish: The birth of the prison* (A. Sheridan, A. Mark, and S. Smith, trans.). New York: Vintage Books.

Frank, R. (1988). Interdisciplinary: The first half-century. In E. G. Stanley & T. F. Hoad (Eds.), *Words: For Robert Burchfield's sixty-fifth birthday* (pp. 91–101). Cambridge, UK: D. S. Brewer.

Freeth, R., Clarke, E., & Fam, D. (2019). Engaging creatively with tension in collaborative research: Harnessing the "I" and "we" through dialogue. In V. Brown, J. Harris, & D. Waltner-Toews. (Eds.), *Independent thinking in an uncertain world: A mind of one's own* (Chapter 15). London and New York: Routledge.

Frey, G. (1973). Methodological problems of interdisciplinary discussions. *Ratio, 15*(1), 161–182.

Frickel, B., Albert, M., & Prainsack, B. (Eds.). (2016). *Investigating interdisciplinary collaboration: Theory and practice across disciplines*. New Brunswick, NJ: Rutgers University Press.

Frickel, S. (2004). *Chemical consequences: Environmental mutagens, scientist activism, and the rise of genetic toxicology*. New Brunswick, NJ: Rutgers University Press.

Friedman, S. S. (2001). Academic feminism and interdisciplinarity. *Feminist Studies*, *27*(2), 499–531.

Friedman, S. S. (2007). Migrations, diasporas, and borders. In D. Nicholls (Ed.), *Introduction to scholarship in modern languages and literatures* (3rd edition) (pp. 209–224). New York: Modern Language Association.

Friman, M. (2010). Understanding boundary work through discourse theory: Inter/disciplines and interdisciplinarity. *Science Studies*, *23*(2), 5–19.

Frodeman, R. (2014). *Sustainable knowledge: A theory of interdisciplinarity*. London: Palgrave Macmillan.

Frodeman, R. (2017). The future of interdisciplinarity: An introduction to the 2nd edition. In R. Frodeman, J. T. Klein, & R. Pacheco (Eds.), *The Oxford handbook of interdisciplinarity* (pp. 3–8). Oxford: Oxford University Press.

Fry, T. (2011). Getting over architecture: Thinking, surmounting, and redirecting transdisciplinary knowledge production. In I. Doucet & N. Janssens (Eds.), *Transdisciplinary knowledge production in architecture and urbanism: Towards hybrid modes of inquiry* (pp. 24–27). Cham, Switzerland: Springer.

Fuller, S. (2017). The military-industrial route to interdisciplinarity. In R. Frodeman, J. T. Klein, & R. Pacheco (Eds.), *The Oxford handbook of interdisciplinarity* (pp. 53–67). Oxford: Oxford University Press.

Funtowicz, S. O. J. R. & Ravetz. (1993). Science for the post-normal age. *Futures*, *25*, 735–755.

Galison, P. (1997). *Image and logic: A material culture of microphysics*. Chicago: University of Chicago Press.

Gardner, H. (2004). *Changing minds: The art and science of changing our own and other people's minds*. Cambridge, MA: Harvard Business Review Press.

Gaziano. E. (1996). Ecological metaphors as boundary work: Innovation and authority in interwar sociology and biology. *American Journal of Sociology*, *101*(4), 874–907.

Gerson, E. M. (2013). Integration of specialties: An institutional and organizational view. *Studies in History and Philosophy of Biological and Biomedical Sciences*, *44*: 515–524.

Getzels, J. W. & Csikszentmihalyi, M. (1964). *The creative vision: A longitudinal study of problem solving in art*. New York: Wiley.

Gibbons, M. (2005). Engagement with the community: The emergence of a new social contact between society and science. Paper presented at the Griffith University Community Engagement Workshop. Queensland, Australia. March.

Gibbons, M. (2008). Why is knowledge translation important? *Grounding the conversation*. FOCUS. Technical Brief No. 21: 1–9.

Gibbons, M., et al. (1994). *The new production of knowledge: The dynamics of science and research in contemporary societies*. Newbury Park, CA: Sage.

Gibbs, P. (Ed.). (2015). *Transdisciplinary professional learning and practice*. Cham, Switzerland and New York: Springer.

Gieryn, T. F. (1983). Boundary work and the demarcation of science from non-science: Strains and interests in professional ideologies of scientists. *American Sociological Review*, *48*(6), 781–795.

Giroux, H. (1992). *Border crossings: Cultural workers and the politics of education*. London and New York: Routledge.

Global Research Council. (2016). *Principles on Interdisciplinarity*. https://www.globalresearchcouncil.org/fileadmin/documents/GRC_Publications/Statement_of_Principles_on_Interdisciplinarity.pdf.

Gleed, A. & Marchant, D. (2016). *Interdisciplinarity: Survey report for the Global Research Council 2016 Annual Meeting*. https://www.globalresearchcouncil.org/fileadmin/documents/GRC_Publications/Interdisciplinarity_Report_for_GRC_DJS_Research.

Golja, T., McClean, S., & Jordan, K. (2018). Collective learning in an industry-education-research test bed. In D. Fam, L. Neuhauser, & P. Gibbs (Eds.), *Transdisciplinary theory, practice, and education: The art of collaborative research and collective learning* (pp. 185–201). Cham, Switzerland and New York: Springer.

Graf, J. (2019). Bringing concepts together: Interdisciplinarity, transdisciplinarity, and SSH integration. *fteval: Journal for Research and Technology Policy Evaluation, 48*, 35–38.

Graff, H. (2015). *Undisciplining knowledge: Interdisciplinarity in the twentieth century.* Baltimore, MD: Johns Hopkins University Press.

Gray, B. (2008). Enhancing transdisciplinary research through collaborative leadership. *American Journal of Preventive Medicine, 35*(2S), S124–S132.

Greenblatt, S. & Gunn, G. (1992). Introduction. In S. Greenblatt & G. Gunn (Eds.), *Redrawing the boundaries: The transformation of English and American literary studies* (pp. 1–11). New York: Modern Language Association.

Gunawardena, S., Weber, R., & Agosto, D. (2010). Finding that special someone: Interdisciplinary collaboration in an academic context. *Journal of Education for Library and Information Science, 51*, 4210–4221.

Gunn, G. (1992). *Thinking across the American grain: Ideology, intellect and the new pragmatism.* Chicago: University of Chicago Press..

Guston, D. H. (2001). Boundary organizations in environmental policy and science: An introduction. *Science, Technology, and Human Values, 26*(4), 399–408.

Habermas, J. (1987). *The theory of communicative action.* Vol. 2: *Lifeworld and system: The critique of functionalist reason* (T. McCarthy, trans.). Boston: Beacon Press.

Hall, K. (2017). What makes teams tick. *Nature*, 551: 562–563.

Hall, K., Vogel, A., Stipelman, B., Stokols, D., Morgan, G. & Gehlert, S. (2012). A four-phase model of transdisciplinary team-based research: Goals, team processes, and strategies. *Translational Behavioral Medicine, 2*(4), 415–430.

Hall, T. E., & O'Rourke, M. (2014). Responding to communication challenges in transdisciplinary sustainability science. In K. Huutoniemi & P. Tapio (Eds.), *Heuristics for transdisciplinary sustainability studies: Solution-oriented approaches to complex problems* (pp. 119–139). London and New York: Routledge.

Hamera, J. (Ed.). (2006). *Opening acts: Performance in/as communication and cultural studies.* Thousand Oaks, CA: Sage.

Haraway, D. (2016). *Staying with the trouble: Making kin in the Chthulucene.* Durham, NC: Duke University Press.

Hardwig, J. (1985). Epistemic dependence. *Journal of Philosophy, 82*, 335–349.

Hartman, J. E. & Messer-Davidow, E. (1991). Introduction: A position statement. In J. E. Hartman & E. Messer-Davidow (Eds.), *(En)Gendering knowledge: Feminists in academe* (pp. 1–7). Knoxville: University of Tennessee Press.

Heite, C. (2012). Setting and crossing boundaries: Professionalization of social work and social work professionalism. *Social World and Society International Online Journal, 10*, online.

Hendren, C. O. & Ku, S. T. (2019). The interdisciplinary executive scientist: Connecting scientific ideas, resources, and people. In K. Hall, A. Vogel, & R. T. Croyle (Eds.), *Strategies for team science success: Handbook of evidence-based principles for cross-disciplinary science and practice lessons learned from health researchers* (pp. 363–373). Cham, Switzerland and New York: Springer.

Hindenlang, K. E., Heeb, J., & Roux, M. (2008). Sustainable coexistence of ungulates and trees: A stakeholder platform for resource use negotiations. In G. Hadorn et al. (Eds.), *Handbook of transdisciplinary research* (pp. 315–326). Cham, Switzerland and New York: Springer.

Hinrich, C. (2008). Sustainable prevention of water associated infection risks: An awareness campaign using visual media. In G. Hadorn et al. (Eds.), *Handbook of transdisciplinary research* (pp. 315–326). Cham, Switzerland and New York: Springer.

Hirsch Hadorn, G.., et al. (Eds.) (2008). *Handbook of transdisciplinary research*. Cham, Switzerland and New York: Springer.

Hirsch Hadorn, G., Bradley, D., Pohl, C., Rist, C., & Wiesmann, U. (2006). Implications of transdisciplinarity for sustainability research. *Ecological Economics*, *60*(1), 119–128.

Holbrook, J. B. (2012). What is interdisciplinary communication? Reflections on the very idea of disciplinary integration. *Synthese*, *190*, 1865–1879.

Holley, K. (2017). Administering interdisciplinary programs. In R. Frodeman, J. T. Klein, & R. Pacheco (Eds.), *The Oxford handbook of interdisciplinarity* (pp. 530–542). Oxford and New York: Oxford University Press.

Huggan, G. (2002). Mixing disciplines: The anxiety of interdisciplinarity. *Postcolonial Studies*, *5*, 256.

Hultén, M. (2013). Boundary objects and curriculum change: The case of integrated versus subject-based teaching. *Journal of Curriculum Studies*, *45*(6), 790–813.

Huutoniemi, K. (2014). Introduction: Sustainability, transdisciplinarity and the complexity of knowing. In K. Huutoniemi & P. Tapio (Eds.), *Transdisciplinary sustainability science: A heuristic approach* (pp. 1–20). London and New York: Routledge.

Huutoniemi, K. & Rafols, I. (2017). Interdisciplinarity in research evaluation. In R. Frodeman, J. T. Klein, & R. Pacheco (Eds.), *The Oxford handbook of interdisciplinarity* (pp. 498–512). Oxford: Oxford University Press.

Imbruce, V. & Prazak, M. (2021). The challenges of studying place: Learning from the failures of an experimental interdisciplinary and community-engaged environmental studies course. In D. Fam & M. O'Rourke (Eds.), *Interdisciplinary and transdisciplinary failures: Lessons learned from cautionary tales* (pp. 181–197). London and New York: Routledge.

INTREPID. (2017). *Recommendations on integrating interdisciplinarity: Interdisciplinarity, the social sciences, and the humanities and responsible research and innovation in EU research*. European Cooperation in Science and Technology.

Jackson, S. (2004). *Professing performance: Theatre in the academy from philosophy to performativity*. Cambridge: Cambridge University Press.

Jacobs, J. A. (2013). *In defense of disciplines: Interdisciplinarity and specialization in the research university*. Chicago: University of Chicago Press.

Jacobs, J. A. & Frickel, S. (2009). Interdisciplinarity: A critical assessment. *Annual Review of Sociology*, *35*, 43–65.

Jahn, T., Bergmann, M., & Keil, F. (2012). Transdisciplinarity: Between mainstreaming and marginalization. *Ecological Economics*, *79*, online.

Jantsch, E. (1972). Towards interdisciplinarity and transdisciplinarity in education and innovation. In L. Apostel et al. (Eds.), *Interdisciplinarity: Problems of teaching and research in universities* (pp. 97–121). Paris: Organization for Economic Cooperation and Development.

Janz, B. (2017). Introduction. In B. Janz (Ed.), *Place, space, and hermeneutics* (pp. 1–5). Cham, Switzerland and New York: Springer.

Jasanoff, S. (2017). A field of its own: The emergence of science and technology studies. In R. Frodeman, J. T. Klein, & R. Pacheco (Eds.), *The Oxford handbook of interdisciplinarity* (pp. 173–187). Oxford: Oxford University Press.

Journet, D. (1993). Interdisciplinary discourse and "boundary rhetoric": The case of S. E. Jelliffe. *Written Communication*, *10*(4), 510–541.

Kann, M. (1979). The political culture of interdisciplinary explanation. *Humanities in Society*, *2*(3), 185–300.

Kannegietter, T. H. (2010). Boundary weaving: The social structure and processes of organizational boundaries. PhD dissertation. Faculty of Science and Technology, Queensland University of Technology.

Kaplan, E. A. & Levine, G. (1997). Introduction. In E. A. Kaplan & G. Levine (Eds.), *The politics of research* (pp. 1–18). New Brunswick, NJ: Rutgers University Press.

Kaplan, S. & Radin, J. (2011). Bounding an emerging technology: Para-scientific media and the Drexler-Smalley debate about nanotechnology. *Social Studies of Science*, *41*(4), 457–485.

Katz, S. (1996). Beyond the disciplines. Address at a meeting on the role of the new American college in the past, present, and future of American higher education.Saint Mary's College of California, Moraga. 17 June.

Keestra, M. (2019). Imagination and actionability: Reflections on the future of interdisciplinarity inspired by Julie Thompson Klein. *Issues in Interdisciplinary Studies*, *37*, 110–129.

Kellner, D. (1995). *Media culture: Cultural studies, identity, and politics: Between the modern and the postmodern*. London and New York: Routledge.

Kessel, F., & Rosenfield, P. (2008). Toward transdisciplinary research: Historical and contemporary perspectives. *American Journal of Preventive Medicine*, *35*, S225–S234.

Keyton, J. (1999). Relational communication in groups. In L. Frey, D. Gouran, & M. S. Poole (Eds.), *The handbook of group communication theory and research* (pp. 192–222). Thousand Oaks, CA: Sage.

Khuri, S. (2018). October 26, 2017: Team science education and training. Intereach, 7 February. https://www.intereach.org/webinars/2018/2/7/october-26-2017-team-science-education-and-training.

Kinchy, A. J. & Kleinman, D. L. (2003). Organizing credibility: Discursive and organizational orthodoxy on the borders of ecology and politics. *Social Studies of Science*, *33*(6), 869–896.

Klein, J. T. (1990). *Interdisciplinarity: History, theory, and practice*. Detroit, MI: Wayne State University Press.

Klein, J. T. (1996). *Crossing boundaries: Knowledge, disciplinarities, and interdisciplinarities*. Charlottesville: University Press of Virginia.

Klein, J. T. (2005). *Humanities, culture, and interdisciplinarity: The changing American academy*. Albany, NY: SUNY Press.

Klein, J. T. (2008). Evaluation of interdisciplinary and transdisciplinary research. *American Journal of Preventive Medicine*, *35*(2S), S116–S123.

Klein, J. T. (2010). *Creating interdisciplinary campus cultures*. San Francisco, CA: Jossey Bass and the Association of American Colleges and Universities.

Klein, J. T. (2013). Communication and collaboration in interdisciplinary research. In M. O'Rourke, S. Crowley, S. Eigenbrode, & J. D. Wulfhorst (Eds.), *Enhancing communication and collaboration in interdisciplinary research* (pp. 11–30). Thousand Oaks, CA: Sage.

Klein, J. T. (2014). Discourses of transdisciplinarity: Looking back to the future. *Futures*, 63: 68–74.

Klein J. T. (2017). Typologies of interdisciplinarity. In R. Frodeman, J. T. Klein, & R. Pacheco (Eds.), *The Oxford handbook of interdisciplinarity* (pp. 21–34). Oxford: Oxford University Press.

Klein, J. T. (2020). Beyond interdisciplinarity: The boundary work of interdisciplinarity. In W. Feuerhahn & R. Mandressi (Eds.), *L'interdisciplinarité: Histoires, politiques et pratiques*. Paris: Éditions de l'Ecole des Haute Etudes en Sciences Sociales. (in French)

Klein, J. T. & Falk-Krzesinski, H. J. (2017). Interdisciplinary and collaborative work: Framing promotion and tenure practices and policies. *Research Policy*, *46*, 1055–1061.

Klein, J. T. & Newell, W. H. (1997). Advancing interdisciplinary studies. In J. G. Gaff & J. L. Ratcliff (Eds.), *Handbook of the undergraduate curriculum: A comprehensive guide to purposes, structures, practices, and change* (pp. 393–415). San Francisco, CA: Jossey-Bass and AACU.

Klein, J. T., Misra, S., & Lotrecchiano, G. (2020). Introduction: Communication in inter- and trans-disciplinary teams. In G. Lotrecchiano & S. Misra (Eds.), *Communication in transdisciplinary teams* (pp. 2–17). Santa Rosa, CA: Informing Science Press.

Klein, et al., Eds. (2001) *Transdisciplinarity*: Joint *problem solving among science technology and society*. Basel: Birkhauser.

Knapp, C. N., Reid, S. R., Fernández-Gimeněz, M. E., Klein, J., & Galvin, K. A. (2019). Placing transdisciplinarity in context: A review of approaches to connect scholars, society, and action. *Sustainability, 11*, 1–25.

Kockelmans, J. J. (1979a). Science and discipline: Some historical and critical reflections. In J. Kockelmans (Ed.), *Interdisciplinarity and higher education* (pp. 11–48). University Park: Pennsylvania State University Press.

Kockelmans, J. J. (1979b). Why interdisciplinarity? In J. J. Kockelmans (Ed.), *Interdisciplinarity and higher education* (pp. 123–160). University Park: Pennsylvania State University Press.

Koestler, A. (1964). *The act of creation*. New York: Penguin Books.

Kolodny, A. (1980). Dancing through the minefield: Some observations on the theory, practice, and politics of a feminist literary criticism. *Feminist Studies, 6*, 1–25.

König, T. & Gorman, M. E. (2017). The challenge of funding interdisciplinary research: A look inside public research funding agencies. In R. Frodeman, J. T. Klein, & R. Pacheco (Eds.), *The Oxford handbook of interdisciplinarity* (pp. 513–524). Oxford: Oxford University Press.

Krohn, W. (2017). Interdisciplinary cases and disciplinary knowledge: Epistemic challenges of interdisciplinary research. In R. Frodeman, J. T. Klein, & R. Pacheco (Eds.), In *The Oxford handbook of interdisciplinarity* (pp. 40–52). Oxford: Oxford University Press.

Kuntz, A. M. & Petrovic, J. E. (2011). The politics of survival in foundations of education: Borderlands, frames, and strategies. *Educational Studies, 47*, 174–197.

Kurian, G. T., et al. (Eds.) (2001). *Encyclopedia of American studies*. New York: Grolier Education, Scholastic Incorporated.

Lambert, R. (1991). Area studies. In D. Easton & C. S. Schelling (Eds.), *Divided knowledge: Across disciplines, across cultures* (pp. 712–732). Thousand Oaks, CA: Sage.

Lamont, M. & Molnár, V. (2002). The study of boundaries in the social sciences. *Annual Review of Sociology, 28*, 167–195.

Lang, D. et al. (2012). Transdisciplinary research in sustainability science: Practices, principles, and challenges. *Sustainability Science*, 7 (Supplement), 25–43.

Last, A. (2018). Of interdisciplinarity. In C. Lury et al. (Eds.), *Routledge handbook of interdisciplinary research methods* (pp. 197–208). London and New York: Routledge.

Lattuca, L. (2001). *Creating interdisciplinarity: Interdisciplinary research and teaching among college and university faculty*. Nashville, TN: Vanderbilt University Press.

Laursen, B. K. (2018). What is collaborative interdisciplinary reasoning: The heart of interdisciplinary team research. In G. Lotrecchiano & S. Misra (Eds.), *Communication in transdisciplinary teams* (pp. 55–80). Santa Rosa, CA: Informing Science Press.

Laursen, B. & O'Rourke, M. (2019). Thinking with Klein about integration. *Issues in Interdisciplinary Studies, 37*(2), 33–61.

Lave, J. & Wenger, E. (1991). *Situated learning: Legitimate peripheral participation*. Cambridge: University of Cambridge Press.

League of European Research Universities. (2014). *Social sciences, humanities, and interdisciplinary research: A showcase of excellent research projects from LERU universities*. Leuven: League of European Research Universities.

Ledford (2008). Collaborations: With all good intentions. *Nature*, 452 (10 April), 682–684.

Leibowitz, B., Wisker, G., & Lamberti, P. (2016). Postgraduate study in uncharted territory. In M. Fourie-Malherbe, C. Aitchison, E. Blitzer, & R. Albertyn (Eds.), *Postgraduate supervision: Future foci for the knowledge society*. Stellenbosch: Sun Press.

Lenz, G. H. (1982). American studies: Beyond the crisis? Recent redefinitions and the meaning of theory, history, and practical criticism. *Prospects: Annual of American Cultural Studies, 7*, 53–113.

Leshner, A. I. (2004). Science at the leading edge. *Science*, 6 February: 729.

Lichnerowicz, A. (1972). Mathematics and transdisciplinarity. In L Apostel et al. (Eds.), *Interdisciplinarity: Problems of teaching and research in universities* (pp. 121–127). Paris: Organization for Economic Cooperation and Development.

Liljegren, A. (2012). Key metaphors in the sociology of professions: Occupations as hierarchies and landscapes. *Comparative Sociology, 11*(1), 88–112.

Limerick, P. (1997). Insiders and outsiders: The borders of the USA and the limits of the ASA. *American Quarterly, 49*(3), 449–469.

Lindvig, K. (2017). Creating interdisciplinarity within monodisciplinary structures. PhD dissertation. University of Copenhagen.

Lionnet, F. & Shih, S. (2011). *The creolization of theory*. Durham, NC: Duke University Press.

Lipsitz, G. (2015). What is American studies? American Studies Association, April. https://www.theasa.net/sites/default/files/What_is_American_Studies.pdf.

Liu, A. (2008). *Local transcendence: Essays on postmodern historicism and the database*. Chicago: University of Chicago Press.

Long, J. C., Cunningham, F. C., & Braithwaite, J. (2013). Bridges, brokers and boundary spanners in collaborative networks: A systematic review. *BMC Health Services Research, 13*, online.

Lotrecchiano, G. & Misra, S. (2020). Transdisciplinary knowledge producing teams: Team processes, knowledge, skills, and competencies. In G. Lotrecchiano & S. Misra (Eds.), *Communication in transdisciplinary teams* (pp. 19–54). Santa Rosa, CA: Informing Science Press.

Lury, C. (2018). Introduction: Activating the present of interdisciplinary methods. In C. Lury et al. (Eds.), *Routledge handbook of interdisciplinary research methods* (pp. 1–25). London and New York: Routledge.

Lury, C., et al. (Eds.). (2018). *Routledge handbook of interdisciplinary research methods*. London and New York: Routledge.

Lyall, A., Bruce, A., Tait, J., & Meagher, L. (2011). *Interdisciplinary research journeys: Practical strategies for capturing creativity*. London: Bloomsbury Academic.

Lyall, C. (2019). *Being an interdisciplinary academic: How institutions shape university careers*. Cham, Switzerland: Palgrave Macmillan Pivot, Springer Nature.

Lykke, N. & Last, A. (2018). Conversation between Angela Last and Nine Lykke. In C. Lury et al. (Eds.), *Routledge handbook of interdisciplinary research methods* (pp. 228–238). London and New York: Routledge.

Lyon, A. (1992). Interdisciplinarity: Giving up territory. *College English, 54*(6), 681–693.

Lyotard, J.-F. (1979). *La condition postmoderne: Rapport sur le savoir*. Paris: Les Éditions de Minuit.

Lyotard, J.-F. (1986). *The postmodern condition: A report on knowledge* (G. Bennington & B. Massumi, trans.). Minneapolis: University of Minnesota Press.

Madison, D. S. & Hamera, J. (Eds.). (2006). *The Sage handbook of performance studies*. Thousand Oaks, CA: Sage.

Mäkinen, E. (2018). Complexity leadership theory and the leaders of transdisciplinary science. *Informing Science: The International Journal of an Emerging Transdiscipline, 21*, online.

Mäkinen, E. (2020). A new paradigm for research organization: Academic leadership in transdisciplinary science teams. In G. R. Lotrecchiano & S. Misra (Eds.), *Communication in transdisciplinary teams* (pp. 123–147). Santa Rosa, CA: Informing Science Press.

Malpas, J. (2018). *Place and experience: A philosophical topography*. 2nd edition. London and New York: Routledge.

Marcovich, A. & Shinn, T. (2011). Where is disciplinarity going? Meeting on the borderland. *Social Science Information, 50*(304), 582–606.

Maton, K. et al. (2006). Community-based interdisciplinary research: Introduction to the special issue. *American Journal of Community Psychology, 38*, 1–7.

McDonald, D., G. Bammer, & Deane, P. (2009). *Research integration using dialogue methods.* Canberra: Australian National University.

McGregor, S. Transdisciplinarity and transpraxis. *Transdisciplinary: Journal of Engineering and Science, 11*, online.

McGregor, S. L. T. (2015). Transdisciplinary pedagogy in higher education: Transdisciplinary learning, learning cycles, and habits of mind. In P. Gibbs (Ed.), *Transdisciplinary higher education: A theoretical basis revealed in practice* (pp. 3–16). Cham, Switzerland: Springer.

McKeon, R. (1979). The uses of rhetoric in a technological age: Architectonic productive arts. In L. F. Bitzer & E. Black (Eds.), *The prospect of rhetoric: Report of the National Development Project* (pp. 44–63). Englewood Cliffs, NJ: Prentice-Hall.

McMillan, J. (2009). Illuminating "transaction spaces" in higher education: Service learning, "boundary work" and "boundary workers." Paper presented at the Knowledge and Curriculum in Higher Education Symposium. University of Cape Town, 29–30 June.

McMillan, J. (2011). What happens when the university meets the community? Service learning, boundary work and boundary workers. *Teaching in Higher Education, 16*(5), 553–564.

McMillan, J., Goodman, S., & Schmid, B. (2016). Illuminating "transaction spaces" in higher education: University-community partnerships and brokering as "boundary work." *Journal of Higher Education Outreach and Engagement, 20*(3), 8–30.

McMurtry, A. (2013). Reframing interdisciplinary and interprofessional collaboration through the lens of collective and sociomaterial theories of learning. *Issues in Interdisciplinary Studies, 31*, 75–98.

McNie, E. C., et al. (2008). Boundary organizations, objects, and agents: Linking knowledge with action in agroforestry watersheds. Joint CID Graduate Student and Research Fellow Working Paper No. 34 and World Agroforesty Centre-ICRAF Working Paper No. 80. Cambridge, MA: Center for International Development, Harvard University.

Medina, N., Báez, L., & Méndez, L. (2018). Collaborative transdisciplinary research in a small institution: Challenges and opportunities. *Informing Science: The International Journal of an Emerging Transdiscipline, 21*, 235–253.

Menand, L. (2001). Undisciplined. *Wilson Quarterly, 25*(4), 51–59.

Menand, L. (2010). *The marketplace of ideas: Reform and resistance in the American university.* New York: Norton.

Messer-Davidow, E. (2017). Situating feminist studies. In R. Frodeman, J. T. Klein, & R. Pacheco (Eds.), *The Oxford handbook of interdisciplinarity* (pp. 214–227). Oxford: Oxford University Press.

Messer-Davidow, E., Shumway, D., & Sylvan, D. (1993). Preface. In E. Messer-Davidow, D. Shumway, & D. Sylvan (Eds.), *Knowledges: Historical and cultural studies in disciplinarity* (pp. vii–viii). Charlottesville: University Press of Virginia.

Meyer, E. & Peukert, D. (2020). Designing a transformative epistemology of the problematic: A perspective for transdisciplinary sustainability research. *Social Epistemology, 34*, 346–356.

Mezirow, J. (2003). Transformative learning as discourse. *Journal of Transformation Education, 1*, 58–63.

Michael, M. (2018). Compromising. In C. Lury et al. (Eds.), *Routledge handbook of interdisciplinary research methods* (pp. 279–283). London and New York: Routledge.

Miller, R. (1982). Varieties of interdisciplinary approaches in the social sciences. *Issues in Integrative Studies, 1*, 1–37.

Miller, R. C. (2020). Interdisciplinarity: Its meaning and consequences. In *Oxford Research Encyclopedia of International Studies*. Oxford: Oxford University Press and International Studies Association, online.

Misra, S., Stokols, D., & Cheng. L. (2015). The transdisciplinary orientation scale: Factor structure and relation to the integrative quality and scope of scientific publications. *Journal of Translational Medicine and Epidemiology, 3*(2), 1042+.

Mitchell, C., Cordell, D., & Fam, D. (2015). Beginning at the end: The outcome spaces framework to guide purposive transdisciplinary research. *Futures, 65*, 76–96.

Mitchell, W. J. T. (2002). Showing seeing: A critique of visual culture. *Journal of Visual Culture, 1*(2), 165–181.

Mittelstrauss, J. (1992). Unity and transdisciplinarity. *Interdisciplinary Science Review, 18*, 153–157.

Mobjörk, M. (2010). Consulting versus participatory transdisciplinarity: A refined classification of transdisciplinary research. *Futures, 42*, 866–873.

Moody, Z. & Darbellay, F. (2018). Studying childhood, children, and their rights: The challenge of interdisciplinarity. *Childhood, 26*(1), 8–21.

Moran, J. (2002). *Interdisciplinarity*. London and New York: Routledge.

Morton, L. W., Eigenbrodge, S. D., & Martin, T. A. (2015). Architectures of adaptive integration in large collaborative projects. *Ecology and Society, 20*(4), online.

Mueller, D. N. (2017). *Network sense: Methods for visualizing a discipline*. Fort Collins, CO: WAC Clearinghouse and Colorado State University Open Press.

Nadim, T. (2018). Haunting seedy connections. In C. Lury et al. (Eds.), *Routledge handbook of interdisciplinary research methods* (pp. 239–247). London and New York: Routledge.

Nash, J. (2008). Transdisciplinary training: Key components and prerequisites for success. *American Journal of Preventive Medicine, 35*(2S), S133–S140.

NASEM (National Academies of Science, Engineering, and Medicine). (2005). *Facilitating Interdisciplinary Research*. Washington, DC: National Academies Press.

NASEM (National Academies of Science, Engineering, and Medicine). (2014). *Convergence: Transdisciplinary integration of life sciences, physical sciences, engineering, and beyond*. Washington, DC: National Academies Press.

NASEM (National Academies of Science, Engineering, and Medicine). (2015). *Enhancing the effectiveness of team science*. Washington, D.C.: National Academies of Science.

NASEM (National Academies of Science, Engineering, and Medicine). (2019). *Fostering the culture of convergence in research: Proceedings of a workshop*. Washington, DC: National Academies Press.

Neuhauser, L., Richardson, D., Mackenzie, S., & Minkler, M. (2007). Advancing transdisciplinary and translational research practice: Issues and models of doctoral education in public health. *Journal of Research Practice, 3*(2), online.

Newell, W. H. (2001). Powerful pedagogies. In B. L. Smith & J. McCann (Eds.), *Reinventing ourselves: Interdisciplinary education, collaborative learning, and experimentation in higher education* (pp. 196–211). Bolton: Anker Press.

Newell, W. H. (2009). The political life cycle of a cluster college: The Western College Program at Miami University. In T. Augsburg & S. Henry (Eds.), *The politics of interdisciplinary studies: Essays on transformations in American undergraduate programs* (pp. 29–50). Jefferson, NC: McFarland.

Newell, W. H. (2013). The state of the field: Interdisciplinary theory. *Issues in Interdisciplinary Studies, 31*, 22–43.

Nicolescu, B. (2002). *Manifesto of transdisciplinarity* (Karen-Claire Voss, trans.). Albany: State University of New York Press.

Nicolescu, B. (2010). Methodology of transdisciplinarity: Levels of reality, logic of the included middle, and complexity. *Transdisciplinary Journal of Engineering and Science, 1*(1), 17–32.

Nowotny, H., Scott, P., & Gibbons, M. (2001). *Re-thinking science: Knowledge and the public in an age of uncertainty*. Cambridge, UK: Polity Press.

Nurius, P. S. & Kemp, S. P. (2019). Individual-level competencies for team collaboration with cross-disciplinary researchers and stakeholders. In K. Hall, A. Vogel, & R. T. Croyle (Eds.), *Strategies for team science success: Handbook of evidence-based principles for cross-disciplinary science and practice lessons learned from health researchers* (pp. 171–187). Cham, Switzerland and New York: Springer.

Oborn, E. & Dawson S. (2010). Learning across communities of practice: An examination of multidisciplinary work. *British Journal of Management, 21,* 843–858.

O'Brien, L., Marzano, M., & White, R. M. (2013). Participatory interdisciplinarity: Towards the integration of disciplinary diversity with stakeholder engagement for new models of knowledge production. *Science and Public Policy, 40*(1), 51–61.

OECD. (2020). Addressing societal challenges using transdisciplinary research. *OECD Science, Technology and Industry Policy Papers,* June, 88. Paris: OECD Publishing.

O'Rourke, Crowley, & Gonnerman. (2016). On the nature of cross-disciplinary integration: A philosophical framework. *Studies in History and Philosophy of Biological and Biomedical Sciences 56*: 62–70.

Osbeck, L. M. & Nersessian, N. J. (2010). Forms of positioning in interdisciplinary science practice and their epistemic effects. *Journal for the Theory of Social Behavior, 40*(2), 136–161.

Ostriker, J. P., Holland, P. W., Kuh, C., & Voytuk, J. A. (Eds). (2009). *A guide to the methodology of the National Research Council Assessment of Doctorate Programs.* Washington, DC: National Academies Press.

Ostriker, J. P. & Kuh, C. V. (Eds.). (2003). *Assessing research-doctorate programs: A methodology study.* Washington, DC: National Academies Press.

Pearce, B. (2021). Learning to fail forwards: Operationalizing productive failure in tackling for complex environmental problems. In D. Fam & M. O'Rourke (Eds.), *Interdisciplinary and transdisciplinary failures: Lessons learned from cautionary tales* (pp. 217–236). London and New York: Routledge.

Peck, J. (1989). There's no place like home? Remapping the topography of German studies. *German Quarterly, 62(2)*, 178–187.

Pfirman, S. (Ed.). (2011). *Interdisciplinary hiring and career development: Guidance for individuals and institutions.* Washington, DC: Council of Environmental Deans and Directors, National Council for Science and the Environment.

Piaget, J. (1972). The epistemology of interdisciplinary relationships. In L. Apostel et al. (Eds.), *Interdisciplinarity: Problems of teaching and research in universities* (pp. 127–139). Paris: Organization for Economic Cooperation and Development.

Pickering, A. (2013). Ontology and antidisciplinarity. In A. Berry & G. Born (Eds.), *Interdisciplinarity: Reconfigurations of the social and natural sciences* (pp. 209–225). London and New York: Routledge.

Piso, Z., O'Rourke, M., & Weathers, K. (2016). Out of the fog: Catalyzing integrative capacity in interdisciplinary research. *Studies in History and Philosophy of Science, 56,* 84–94.

Pohl, C., Fam, D., Hoffmann, S., & Mitchell, C. (2019). Exploring Julie Thompson Klein's framework for analysis of boundary work. *Issues in Interdisciplinary Studies, 37*(2), 62–89.

Pohl, C., van Kerkhoff, L., Hadorn, G. H., & Bammer, G. (2008). Integration. In G. H. Hadorn et al. (Eds.), *Handbook of transdisciplinary research* (pp. 411–424). Cham, Switzerland and New York: Springer.

Pohl, C., et al. (2010). Researchers' roles in knowledge co-production: Experience from sustainability research in Kenya, Switzerland, Bolivia and Nepal. *Science and Public Policy, 37*(4), 267–281.

Polk, M. (2015a). Becoming subversive: Challenges, weaknesses, and opportunities in transdisciplinary problem solving research. Keynote address at the td-net Conference on Health and Sustainability. Basel, Switzerland. 8 September.

Polk, M. (2015b). Transdisciplinary co-production: Designing and testing a transdisciplinary research framework for societal problem solving. *Futures, 65,* 110–122.

Popa, F., Guillermin, M., & Dedeurwaerdere, T. (2015). A pragmatist approach to transdisciplinarity in sustainability. *Futures, 65,* 45–56.

Quick, K. S. & Feldman, M. S. (2014). Boundaries as junctures: Collaborative boundary work for building efficient resilience. *Journal of Public Administration Research and Theory, 24,* 673–695.

Ratcheva, V. (2009). Integrating diverse knowledge through boundary spanning processes: The case of multidisciplinary project teams. *International Journal of Project Management, 27*, 206–215.

Readings, B. (1996). *The university in ruins*. Cambridge, MA: Harvard University Press.

Repko, A. (2008). *Interdisciplinary research: Process and theory*. Thousand Oaks, CA: Sage.

Reynolds, B. & Fitzpatrick, J. (1999). The transversality of Michel de Certeau: Foucault's panoptic discourse and the cartographic impulse. *Diacritics, 29*(3), 63–80.

Rhoten, D. (2004). Interdisciplinary research: Trend or transition. *Items & Issues 5*(1–2), 6–11.

Rhoten, D., Boix-Mansilla, V., Chun, M., & Klein, J. T. (2006). *Interdisciplinary education at liberal arts institutions*. Teagle Foundation White Paper. Brooklyn, NY: Social Science Research Council.

Riesch, H. (2010). Theorizing boundary work as representation and identity. *Journal for the Theory of Social Behavior, 24*(1), 452–473.

Robinson, J. (2010). On beyond zebra: Being undisciplined in support of sustainability. In *The Trudeau Foundation Papers* (vol. 2, pp. 96–129). Montreal: Pierre Elliot Trudeau Foundation.

Roco, M. C., Bainbridge, W. S., Toon, B., & Whitesides, G. (2013). Executive summary. In M. C. Roco, W. S. Bainbridge, B. Toon, & G. Whitesides (Eds.), *Convergence of knowledge, technology, and society: Beyond convergence of nano-bio-info-cognitive technologies* (pp. i–lvi). Cham, Switzerland and New York: Springer.

Rosenfeld, P. & Kessel, F. (2008). Toward transdisciplinary research: Historical and contemporary perspectives. *American Journal of Preventive Medicine, 35*(2 Suppl), S225–S234.

Rossini, F. A. & Porter, A. L. (1979). Frameworks for integrating disciplinary research. *Research Policy, 8*, 70–79.

Roy, R. (Ed.) (2000). *The interdisciplinary imperative: Interactive research and education, still an elusive goal in academia*. Lincoln, NE: Writer's Club Press.

Russell, R. A., Wickson, F., & Carew, A. (2008). Transdisciplinarity: Context, contradictions and capacity. *Futures, 40*, 460–472.

Rylance, R. (2015). Global funders to focus on interdisciplinarity. *Nature, 525*(17), 313–315.

Salazar, M. R. & Lant, T. K. (2018). Facilitating innovation in interdisciplinary teams: The role of leaders and integrative communication. *Informing Science: The International Journal of an Emerging Transdiscipline, 21*, 157–178.

Salazar, M. R., Lant, T., Fiore, S., & Salas, E. (2012). Facilitating innovation in diverse science teams through integrative capacity. *Small Group Research, 43*(5), 527–558.

Schauppenlehner-Kloyber, E. & Penker, M. (2015). Managing group processes in transdisciplinary future studies. *Futures, 65*, 57–71.

Schleifer, R. (2002). A new kind of work: Publishing, theory, and cultural studies. In D. Shumway & C. Dionne (Eds.), *Disciplining English* (pp. 179–194). Albany, NY: SUNY Press.

Schmithals, J. & Berkenhagen, J. (2004). Beziehungen und motivation. In M. Schophaus, S. Schön, & H.-L. Dienel (Eds.), *Transdisziplinäres looperaationsmanagement* (pp. 69–102). Munich: Ökom Verlag.

Scholle, D. (1995). Resisting disciplines: Repositioning media studies in the university. *Communication Theory, 5*, 130–143.

Schön, D. (1984). *The reflective practitioner: How professionals think in action*. New York: Basic Books.

Schor, N. (1992). Feminist and gender studies. In J. Gibaldi (Ed.), *Introduction to scholarship in modern languages and literatures* (pp. 262–287). New York: Modern Language Association.

Scholz, R. W., et al. (2011). *Environmental literacy in science and society*. Cambridge: Cambridge University Press.

Seabury, M. B. (1999). Introduction. In M. B. Seabury (Ed.), *Interdisciplinary general education: Questioning outside the lines* (pp. 1–25). New York: College Board.

Sharp, P., et al. (2011). *The third revolution: The convergence of the life sciences, physical sciences, and engineering*. Cambridge, MA: MIT Press.

Sharp, P. A. & Langer, R. (2011). Promoting convergence in biomedical science. *Science, 333*(6042), 527.

Shinn, T. (2001). A fresh look at instrumentation: An introduction. In B. Joerges & T. Shinn (Eds.), *Instrumentation between science, state, and industry* (pp. 1–13). Dordrecht: Kluwer Academic.

Sill, D. J. (1996). Integrative thinking, synthesis, and creativity in interdisciplinary studies. *Journal of General Education, 45*(2), 129–151.

Smelser, N. J. (2004). Interdisciplinarity in theory and practice. In C. Camis & J. Joas (Eds.), *The dialogical turn: New roles for sociology in the postdisciplinary age* (pp. 43–64). Lanham, MD: Rowman and Littlefield.

Soja, E. W. (1989). *Postmodern geographies: The reassertion of space in critical social theory.* London: Verso.

Spelt, E., Biemans, H., Tobi, H., Luning, P., & Mulder, M. (2009). Teaching and learning in interdisciplinary higher education: A systematic review. *Educational Psychology Review, 21*, 365–378.

Star, S., & Griesemer, J. (1989). Institutional ecology, "transitions" and boundary objects: Amateurs and professionals in Berkeley's Museum of Vertebrate Zoology, 1907–39. *Social Studies of Science, 19*(3), 387–420.

Stimpson, C. R. (1992). Feminist criticism. In Greenblatt and Gunn (Eds.), *Redrawing the boundaries: The transformation of English and American literary studies* (pp. 251–270). New York: Modern Language Association.

Stokols, D. (2017). *Social ecology in the digital age: Solving complex problems in a globalized world.* Waltham, MA: Academic Press Books, Elsevier.

Stone, A. R. (1969). The interdisciplinary research team. *Journal of Applied Behavioral Science, 5*(3), 351–365.

Streck, D. R. (2007). Research and social transformation: Notes about method and methodology in participatory research. *International Journal of Action Research, 3*(1–2): 112–130.

Strober, M. H. (2011). *Interdisciplinary conversations: Challenging habits of thought.* Stanford, CA: Stanford University Press.

Szostak, R. (2017). Book based on "misguided assumptions." *Integrative Pathways, 39*(1), 8–10.

Tebes, J. K. & Thai, N. D. (2018). Interdisciplinary team science and the public: Steps toward a participatory team science. *American Psychology, 73*(4), 549–562.

Thagard, P. (2017). Cognitive science. In R. Frodeman, J. T. Klein, & R. Pacheco (Eds.), *The Oxford handbook of interdisciplinarity* (pp. 188–200). Oxford: Oxford University Press.

Thompson, J. L. (2009). Building collective communication competence in interdisciplinary research teams. *Journal of Applied Communication Research, 37*(3), 278–297.

Tilghman, B. (2006). Crossing boundaries. *British Journal of Aesthetics, 46*(2), 178–191.

Toombs, W. & Tierney, W. (1991). *Meeting the mandate: Reviewing the college and department curriculum.* ASHE-ERIC Higher Education Report No. 6. Washington, DC: School of Educational and Human Development, George Washington University.

Tosey, P., Visser, M., & Saunders, N. K. (2011). The origins and conceptualizations of "triple-loop" learning: A critical review. *Management Learning, 43*(3), 291–307.

Trowler, P. R. & Knight, P. T. (2002). Exploring the implementation gap: Theory and practices in change interventions. In P. R. Trowler (Ed.), *Higher education policy and institutional change: Intentions and outcomes in turbulent environments* (pp. 142–163). Buckingham, UK: Society for Research into Higher Education.

Tuomi-Gröhn, T. & Engeström, Y. (Eds.). (2003). *Between school and work: New perspectives on transfer and boundary-crossing.* Amsterdam: Pergamon.

Turner, B. (1990). The interdisciplinary curriculum: From social medicine to post-modernism. *Sociology of Health and Illness, 12*, 1–23.

Turner, B. L. (2002). Geography: Response to Thrift's "The Future of Geography." *GeoForum*, *33*, 427–429.

Turner, S. (2017). Knowledge formations: An analytic framework. In R. Frodeman, J. T. Klein, & R. Pacheco (Eds.), *The Oxford handbook of interdisciplinarity* (pp. 9–20). Oxford: Oxford University Press.

Ullman, S. (1962). *Semantics: An introduction to the science of meaning*. Oxford: Blackwell.

Utzkowitz, M. (2021). Transdisciplinary research on the way to create a culture of internationality. In D. Fam & M. O'Rourke (Eds.), *Interdisciplinary and transdisciplinary failures: Lessons learned from cautionary tales* (pp. 165–178). London and New York: Routledge.

Van Breda, J., Musango, J., & Brent, A. (2016). Undertaking individual transdisciplinary PhD research for sustainable development: Case studies from South Africa. *International Journal of Sustainability in Higher Education*, *17*(2), 150–166.

Van Dijk, T. A. (2011). Introduction: The study of discourse. In T. A. Van Dijk (Ed.), *Discourse studies: A multidisciplinary introduction* (2nd edition) (pp. 1–7). London: Sage

Vickers, J. (1997). "Unframed in open, unmapped fields": Teaching and the practice of interdisciplinarity. *Arachne: An Interdisciplinary Journal of the Humanities*, *4*, 11–14.

Vickers, J. (2003). Diversity, globalization, and "growing up digital": Navigating interdisciplinarity in the twenty-first century. *History of Intellectual Culture*, *13*, online.

Vienni Baptista, B. (2016). Introduction: Interdisciplinarity in Latin America: Building dialogue through regionalism. *Issues in Interdisciplinary Studies*, *34*, 109–121.

Vienni Baptista, B., Lyall, C., Ohlmeyer, D., Spaapen, J., Wallace, D., & Pohl, C. (2020). *Final report on understanding of interdisciplinary and transdisciplinary research: Policy brief*. SHAPE-ID. https://www.shapeid.eu

Vienni Baptista, B., Mazzitelli, M. G., & Came, F. F. (2020). Transdisciplinary communication in research teams: Institutional constructs and practices from a Uruguayan perspective. In G. R. Lotrecchiano & S. Misra (Eds.), *Communication in transdisciplinary teams* (pp. 253–291). Santa Rosa, CA: Informing Science Press.

Vilsmaier, U. & Thalheimer, A. (2021). Failing and the perception of failure in student-driven transdisciplinary projects. In D. Fam & M. O'Rourke (Eds.), *Interdisciplinary and transdisciplinary failures: Lessons learned from cautionary tales* (pp. 237–252). London and New York: Routledge.

Vogel, A., et al. (2012). Influence of a National Cancer Institute Transdisciplinary Research and Training Initiative on trainees' transdisciplinary research competencies and scholarly productivity. *Translational Behavioral Medicine*, *2*(4), 459–468.

Volet, S., Vaurus, M., & Salonen, P. (2009). Self- and social regulation in learning contexts: An integrative perspective. *Educational Psychologist*, *44*(4), 215–226.

Von Wehrden, H. et al. (2018). Interdisciplinary and transdisciplinary research: Finding the common ground of multi-faceted concepts. *Sustainability Science*, *14*, 875–888.

Vosskamp, W. (1994). Crossing of boundaries: Interdisciplinarity as an opportunity for universities in the 1990s. *Issues in Interdisciplinary Studies*, *12*, 43–54.

Wackerhausen, S. (2009). Collaboration, professional identity and reflection across boundaries. *Journal of Interprofessional Care*, *23*(5), 455–473.

Walker, D. & Nocon, H. (2007). Boundary-crossing competence: Theoretical considerations and educational design. *Mind, Culture, and Activity*, *14*(3), 178–195.

Wallace, R. L. & Clark, S. G. (2017). Barriers to interdisciplinarity in environmental studies: A case of alarming trends in faculty and programmatic wellbeing. *Issues in Interdisciplinary Studies*, *35*, 221–247.

Warhol, R. (2002). "Nice work if you can get it—and if you can't?" Building women's studies without tenure lines. In R. Wiegman (Ed.), *Women's studies on its own: A next wave reader in institutional change* (pp. 224–232). Durham, NC: Duke University Press.

Wátzlawick, P., Weakland, J. H., & Fisch, R. (1974). *Change: Principles of problem formation and problem resolution*. New York: Norton.

Weerts, D. J. & Sandmann, L. R. (2010). Community engagement and boundary-spanning roles at research universities. *Journal of Higher Education, 81*(6), 702–727.

Weinberger, A., Stegmann, K., & Fischer, F. (2007). Knowledge convergence in collaborative learning: Concepts and assessment. *Learning and Instruction, 17*, 416–426.

Weingart, P. (2000). Interdisciplinarity: The paradoxical discourse. In P. Weingart & N. Stehr (Eds.), *Practicing interdisciplinarity* (pp. 25–41). Toronto: University of Toronto Press.

Weingart, P. (2010). A short history of knowledge formations. In R. Frodeman, J. T. Klein, & C. Mitcham (Eds.), *The Oxford handbook of interdisciplinarity* (pp. 3–14). New York: Oxford University Press.

Welch, J. (2011). The emergence of interdisciplinarity from epistemological thought. *Issues in Interdisciplinary Studies, 29*, 1–39.

Wenger, E. (1998). Communities of practice: Learning, meaning, and identity. *Journal of Mathematics Teacher Education, 6*(2), 185–194.

Weszkalnys, G. & Barry, A. (2013). Multiple environments: Accountability, integration and ontology. In A. Berry & G. Born (Eds.), *Interdisciplinarity: Reconfigurations of the social and natural sciences* (pp. 178–208). London and New York: Routledge.

Whatmore, S. J. (2013). Where natural and social science meet? Reflections on an experiment in geographical practice. In A. Berry & G. Born (Eds.), *Interdisciplinarity: Reconfigurations of the social and natural sciences* (pp. 161–177). London and New York: Routledge.

Wiegman, R. (2002). Introduction: On location. In R. Wiegman (Ed.), *Women's studies on its own: A next wave reader in institutional change.* (pp. 1–44). Durham, NC: Duke University Press.

Wiek, A. (2007). Challenges of transdisciplinary research as interactive knowledge generation: Experiences from transdisciplinary case study research. *GAIA, 16*(1), 52–57.

Winter, M. (1996). Specialization, territoriality, and jurisdiction in librarianship. *Library Trends, 45*(2), 343–363.

Wolosky, S. (2013). Preface: Public and private in feminist theory and poetics. In S. Wolosky (Ed.), *Feminist theory across disciplines: Feminist community and American women's poetry* (pp. xi+). London and New York: Routledge.

Woolf, S. H. (2008). The meaning of translation research and why it matters. *JAMA, 299*(2), 211–213.

Yanow, D. (2004). Translating local knowledge at organizational peripheries. *British Journal of Management, 15*(S2), S9–S25.

Zierhofer, W. & Burger, P. (2007). Disentangling transdisciplinarity: An analysis of knowledge integration in problem-oriented research. *Science Studies, 20*(1), 51–74.

Zietsma, C. & Lawrence, T. (2010). Institutional work in the transformation of an organizational field: The interplay of boundary work and practice work. *Administrative Science Quarterly, 55*, 189–221.

Index

For the benefit of digital users, indexed terms that span two pages (e.g., 52–53) may, on occasion, appear on only one of those pages.

The letter *f* following a page locator denotes a figure while the letter *t* denotes a table.

www.ingramcontent.com/pod-product-compliance
Ingram Content Group UK Ltd.
Pitfield, Milton Keynes, MK11 3LW, UK
UKHW020923170926
475595UK00031B/35